Big Data Management

The big data paradigm presents a number of challenges for university curricula on big data or data science related topics. On the one hand, new research, tools and technologies are currently being developed to harness the increasingly large quantities of data being generated within our society. On the other, big data curricula at universities are still based on the computer science knowledge systems established in the 1960s and 70s. The gap between the theories and applications is becoming larger, as a result of which current education programs cannot meet the industry's demands for big data talents.

This series aims to refresh and complement the theory and knowledge framework for data management and analytics, reflect the latest research and applications in big data, and highlight key computational tools and techniques currently in development. Its goal is to publish a broad range of textbooks, research monographs, and edited volumes that will:

- Present a systematic and comprehensive knowledge structure for big data and data science research and education
- Supply lectures on big data and data science education with timely and practical reference materials to be used in courses
- Provide introductory and advanced instructional and reference material for students and professionals in computational science and big data
- Familiarize researchers with the latest discoveries and resources they need to advance the field
- Offer assistance to interdisciplinary researchers and practitioners seeking to learn more about big data.

The scope of the series includes, but is not limited to, titles in the areas of database management, data mining, data analytics, search engines, data integration, NLP, knowledge graphs, information retrieval, social networks, etc. Other relevant topics will also be considered.

Jingyuan Wang · Jiahao Ji

Mechanism-Driven Explainable Urban Spatio-Temporal Prediction

Springer

Jingyuan Wang
School of Computer Science
and Engineering
Beihang University
Beijing, China

Jiahao Ji
School of Computer Science
and Engineering
Beihang University
Beijing, China

ISSN 2522-0179 ISSN 2522-0187 (electronic)
Big Data Management
ISBN 978-981-92-0661-2 ISBN 978-981-92-0662-9 (eBook)
https://doi.org/10.1007/978-981-92-0662-9

This Springer imprint is published by the registered company Springer Nature Singapore Pte Ltd.
The registered company address is: 152 Beach Road, #21-01/04 Gateway East, Singapore 189721, Singapore

Preface

In recent years, the rapid advancement of sensing technologies and information infrastructures has led to the generation of massive volumes of spatio-temporal data within urban systems. Leveraging these data for prediction and decision-making has become an important theme in artificial intelligence and urban computing. In this context, data-driven approaches—particularly deep learning models—have achieved remarkable success in tasks such as urban traffic forecasting. Nevertheless, most mainstream studies largely overlook the underlying mechanisms through which urban systems generate spatio-temporal data. They usually simplify complex urban dynamics into black-box mappings that can be automatically fitted in an end-to-end manner. While this paradigm has delivered impressive predictive performance, it has also revealed several fundamental limitations: limited interpretability, sensitivity to scenario changes and distribution shifts, and a lack of insight into intrinsic mechanisms and causal structures of urban systems.

The motivation for this book arises from a critical reflection on this prevailing paradigm. We argue that urban spatio-temporal systems are, in essence, complex dynamical systems shaped jointly by behavioral mechanisms and environmental factors. Their evolution cannot be adequately characterized by a single homogeneous structure. Purely data-driven statistical fitting, therefore, is insufficient to uncover the mechanisms that govern such systems. Instead, constructing a framework that integrates mechanism-driven and data-driven approaches, and incorporating theoretical knowledge, physical constraints, and causal structures into model design may offer a more interpretable and potentially more generalizable path forward.

Guided by this central idea, the book systematically investigates mechanism-driven modeling for urban spatio-temporal prediction from the dual perspectives of intrinsic behavioral factors and extrinsic contextual influences. We begin by reviewing the historical development and dominant paradigms in urban spatio-temporal forecasting, analyzing their theoretical assumptions and inherent limitations. Building upon this foundation, the book presents a progressive exploration spanning mechanism modeling, neural architecture design, causal representation learning, and loss function construction. Finally, we discuss the broader implications of the "mechanism-data" integration paradigm and outline promising future directions. Rather than proposing a universal model, this book emphasizes a modeling philosophy: *how to embed mechanistic principles within deep learning frameworks so that the predictive process and outcomes carry clearer physical meaning and stronger generalization capability.*

This book is intended for researchers and graduate students in artificial intelligence, urban computing, spatio-temporal data mining, and complex systems modeling. It also seeks to provide insights for scholars interested in model interpretability and the integration of mechanism-driven and data-driven paradigms. For engineering practitioners, the frameworks presented herein may offer inspiration for addressing spatio-temporal prediction problems in complex real-world systems.

The complexity of urban systems implies that any model can only approximate reality. The framework proposed in this book represents one step toward mechanism-driven and interpretable artificial intelligence. We hope that integrating mechanistic thinking with deep learning will offer new perspectives for predicting and understanding complex spatio-temporal systems in the years to come.

Beijing, China
February 2026

Jingyuan Wang
Jiahao Ji

Acknowledgements The book was supported by the National Natural Science Foundation of China (No. 72222022, 72242101, 72171013). The completion of this book would not have been possible without the support and contributions of many individuals. We would like to express our sincere gratitude to all who contributed to its development. In particular, we thank Qian Lei, Zinan Liu, Ying Pang, and Zhenghao Zhou (listed in alphabetical order by surname) for their dedicated efforts in material organization and proofreading. The research presented in this book is the result of long-term academic accumulation and collaborative exchange. We are deeply grateful to all collaborators and colleagues who provided valuable insights and support.

Competing Interests The authors have no competing interests to declare that are relevant to the content of this manuscript.

Contents

Chapter 1
Introduction to Urban Spatio-Temporal Prediction

Abstract This chapter introduces the background, motivation, and overarching vision of the book. As modern cities increasingly rely on data-driven intelligence, spatio-temporal prediction has become a fundamental tool for understanding and managing complex urban systems, including transportation, environmental monitoring, and public safety. Despite recent advances in deep learning, many existing approaches remain limited by their black-box nature and sensitivity to distribution shifts, which restricts their reliability in real-world decision-making scenarios. To address these challenges, this chapter outlines the central premise of the book: integrating domain mechanisms with modern learning models to achieve interpretable, robust, and generalizable spatio-temporal prediction. It provides a high-level overview of the current research landscape, highlighting the strengths and limitations of mechanism-based models, data-driven deep learning, and emerging hybrid approaches. The chapter further clarifies the scope and objectives of the book, explaining how each subsequent chapter contributes to a unified framework for mechanism-driven modeling. Finally, it offers a roadmap of the book's structure to guide readers through the progression from foundational concepts to advanced modeling paradigms and real-world applications.

1.1 Background and Motivation

Driven by the continuous improvement of urban infrastructure and the widespread application of Internet of Things (IoT) technology, various fields closely related to human life are constantly accumulating massive amounts of spatio-temporal data [2]. Such data not only contains the information it carries but also has temporal and spatial attributes, describing the time point and geographical location where the information occurs—hence the name *spatio-temporal data* [79].

Typical examples of spatiotemporal data[1] include mobile phone signaling data generated by residents' daily travel, Point of Interest (POI) check-in data, real-time GPS positioning data recorded during the operation of various vehicles, passenger

[1] The terms "spatiotemporal" and "spatio-temporal" are used interchangeably in this book.

J. Wang and J. Ji, *Mechanism-Driven Explainable Urban Spatio-Temporal Prediction*, Big Data Management, https://doi.org/10.1007/978-981-92-0662-9_1

flow data of public transportation tools, temporal dynamic statistics of different populations during the spread of infectious diseases, and real-time meteorological and pollutant data collected by meteorological stations. These data are continuously accumulated in daily life and show an exponential growth trend [1].

Taking the transportation field as an example, DiDi Chuxing platform[2] generates an average of more than 33 million vehicle trajectory data points per day; as of January 2023, the daily positioning calls of BeiDou Navigation Satellite System (BDS)[3] on the Amap[4] mobile application have exceeded 300 billion times, with each call generating a piece of spatiotemporal data. These large-scale and high-frequency updated data not only reflect the spatiotemporal patterns of individual and group behavior but also contain rich information, providing unprecedented data support for revealing the operating laws of complex urban systems [55].

Spatio-temporal prediction aims to accurately forecast future changes of spatio-temporal data by learning dependencies within historical observations. It enables the prediction of system evolutionary trends and supports scientific decision-making. This technology has become a research hotspot in academia and industry, attracting the attention of many researchers [22, 60]. According to statistics from the Web of Science platform, the number of academic papers in this field has shown explosive growth over the past decade (see Fig. 1.1a).

Existing studies mainly focus on improving the accuracy of prediction models in complex scenarios to meet the needs of refined urban governance. For example, modeling spatio-temporal traffic flow data and predicting future trends helps alleviate traffic congestion, optimize signal control strategies, and improve travel efficiency. The analysis and prediction of spatio-temporal trajectories of hazardous chemical vehicles can identify potential explosion risk points and other safety hazards in advance. This assists relevant departments in taking preventive measures and enhances urban safety [20, 54, 56]. In addition, in the early stage of an epidemic, predicting the geographical distribution and transmission speed of infected individuals can effectively guide the formulation of isolation strategies and the allocation of medical resources [57–59].

The rapid development of Artificial Intelligence (AI) technology, especially the breakthrough progress of deep learning, has provided brand-new technical support for spatio-temporal prediction. By constructing Deep Neural Networks (DNNs), deep learning can automatically extract complex features from large-scale data, demonstrating strong nonlinear modeling capabilities [26]. In terms of spatio-temporal data modeling, methods such as Recurrent Neural Networks (RNNs) and their variants [7, 18, 19] as well as attention mechanisms [52], which target temporal dependencies, have significantly improved the expressive power of time series prediction. For spatial

---

[2] A Chinese mobility technology company that offers application-based transportation and related services, including ride-hailing, taxi services, bike sharing, and vehicle leasing. See https://en.wikipedia.org/wiki/DiDi for details.

[3] The BDS is a satellite-based radio navigation system owned and operated by the China National Space Administration. See https://en.wikipedia.org/wiki/BeiDou for details.

[4] A web mapping, navigation, and location-based services provider similar to Google Maps.

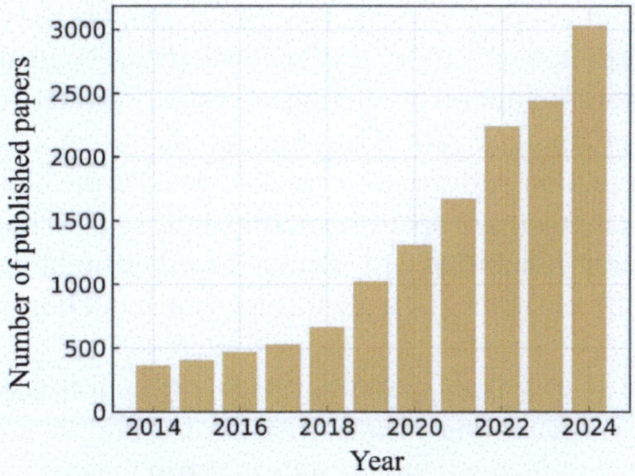

(a) Number of papers grows exponentially

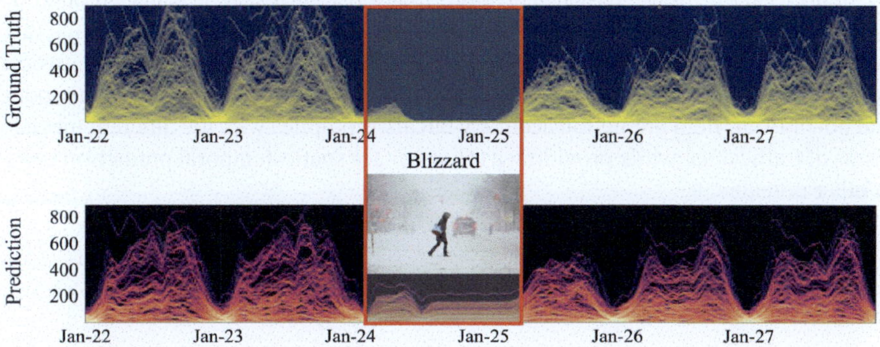

(b) Prediction performance significantly declines under extreme weather scenarios [?]

Fig. 1.1 Motivation

relationships, Convolutional Neural Networks (CNNs) [27] and Graph Neural Networks (GNNs) [63] have provided effective solutions for modeling complex spatial networks.

Driven by this, top universities and research institutions worldwide have launched research on deep learning in the field of spatio-temporal prediction for urban scenarios in recent years. For example, the Senseable City Lab at the Massachusetts Institute of Technology (MIT) in the United States and the Urban Lab at University College London (UCL) in the United Kingdom have achieved important research results in application scenarios such as intelligent transportation systems and urban travel optimization.

Although deep learning has achieved remarkable results on benchmark datasets in the field of spatio-temporal prediction, it still faces numerous challenges in practical urban management scenarios. *On one hand*, the prediction performance of existing data-driven methods degrades significantly under abnormal conditions such as extreme weather and sudden accidents [61], exposing their limitations in modeling the mechanisms of complex systems (see Fig. 1.1b). *On the other hand*, unlike other deep learning applications (e.g., text analysis [66, 67], image processing [16, 75]), the results of spatio-temporal prediction are usually directly used to support the decision-making of urban managers. Therefore, prediction models not only need to achieve excellent prediction performance but also possess the ability to interpret the internal evolution mechanisms of spatio-temporal data, so as to improve model interpretability and decision reliability. This demand further highlights the importance of integrating data-driven methods with mechanism modeling in spatio-temporal prediction.

Traffic prediction is a typical spatio-temporal prediction task. Due to the significant spatio-temporal coupling characteristics of traffic data and the important decision-making value of its prediction results in urban management and traffic regulation, traffic prediction occupies a core position in the field of spatio-temporal prediction. Studies have shown that more than 60% of research papers related to spatio-temporal prediction focus on traffic prediction tasks [22]. Based on this, this book takes traffic prediction as the entry point, focuses on the mechanism fusion problem of deep prediction models, and aims to construct a traffic prediction model that not only has high prediction accuracy but also complies with the inherent mechanisms of traffic data, while providing a reference for spatio-temporal prediction tasks in other domains.

1.2 Significance and Relevance of the Topic

This section highlights the theoretical and practical significance of mechanism-driven explainable spatio-temporal prediction. By bridging mechanism-based modeling and data-driven deep learning, the book advances a hybrid paradigm that enriches the theoretical foundations of spatio-temporal prediction and promotes interpretable, generalizable models. Beyond theory, the proposed methods address real-world challenges in urban governance and digital platforms, supporting reliable decision-making in transportation, public health, and intelligent urban services.

1.2.1 Theoretical Significance

Spatio-temporal prediction is one of the important frontier research directions in the field of artificial intelligence. Considering the limitations of existing deep spatio-temporal prediction methods and the characteristics of mechanism models, this book proposes a mechanism-driven, interpretable deep spatio-temporal prediction method to address the shortcomings of existing approaches. This research not only enriches the theoretical system in the field of spatio-temporal prediction but also provides new ideas and methodologies for research in other interdisciplinary fields. Specifically, this book has the following theoretical significance.

(1) Enriching the theoretical system of spatio-temporal prediction. Existing spatio-temporal prediction research mainly focuses on the data-driven paradigm, but models built based on this paradigm lack the ability to model the internal mechanisms of the system. This book introduces a series of deep spatiotemporal prediction methods that integrate mechanism models, exploring a novel modeling paradigm that combines mechanism-driven and data-driven approaches, thus expanding the research boundaries of the spatiotemporal prediction field. This theoretical innovation can not only improve the adaptability of prediction models to complex urban scenarios but also provide more comprehensive methodological support for spatiotemporal data modeling.

(2) Promoting the development of hybrid models that combine mechanism-driven and data-driven approaches. This book proposes a systematic paradigm for integrating mechanism-driven and data-driven models. Especially in urban transportation systems, technological breakthroughs in issues such as how to integrate mechanism models corresponding to behavioral and data factors into the deep learning modeling process will not only further deepen human's understanding of the relationship between data and physical laws in complex systems, but will also prompt researchers in this field to pay more attention to the mining and modeling of mechanisms when developing data-driven models, thereby promoting the development of "mechanism-data" hybrid driven models.

(3) Providing insights and references for research in other interdisciplinary fields. The research idea of combining deep learning with mechanism modeling to address issues such as insufficient physical consistency and poor generalization ability of deep models is not only applicable to spatio-temporal prediction but also can be extended to other interdisciplinary fields, such as bioinformatics and medical auxiliary decision-making. It introduces the "white-box" mechanisms of the field into the construction of "black-box" deep models, designing a "gray-box" model that can produce mechanism-compliant results in complex scenarios. Therefore, this book has broad theoretical reference significance and provides new ideas for solving similar problems.

1.2.2 Practical Significance

Spatio-temporal prediction is not only one of the cutting-edge academic research directions in AI but also a field derived from practical application problems, developed through academic research, and required to return to practical application scenarios. Therefore, it holds important management and practical significance in social governance scenarios such as urban traffic management, public health emergencies, and urban safety emergency response, as well as in commercial application fields including map platforms, shared mobility, and food delivery. Specifically, this book has the following practical significance.

(1) Contributing to the modernization and refinement of urban governance. The deep spatiotemporal prediction methods based on the integration mechanism model

proposed in this book can be applied to multiple important fields of urban management.

For example, in urban traffic management, high-precision prediction of traffic flow and congestion trends can optimize traffic signal control strategies, alleviate urban congestion, and improve the level of smart transportation construction. In public health, governments and public health institutions can use spatiotemporal prediction models to predict the development trend of epidemics in advance, thereby formulating prevention policies, optimizing resource allocation, and promoting the upgrading of public health prevention and control paradigms. In the field of safety supervision of hazardous chemical transportation, spatiotemporal prediction technology can be used to accurately simulate the operating trajectory of hazardous chemical transport vehicles, provide early warnings in high-risk areas, identify potential safety hazards in advance, ensure urban safety, and promote the intellectualization of urban safety and emergency response systems. The implementation of these application scenarios will effectively enhance the modernization and refinement of urban governance.

(2) Improving the service quality of relevant industry application platforms. People's daily lives are inseparable from the support of software such as map platforms, ride-sharing, and food delivery, and one of the core technologies behind these software programs is spatiotemporal prediction. The method introduced in this book can not only enhance the spatiotemporal prediction effect of these software programs but also ensure their service quality in unforeseen scenarios, thereby improving user experience, increasing user stickiness, and promoting the business development of companies operating related industry application platforms.

For example, in ride-sharing platforms, providing accurate spatiotemporal supply and demand predictions for specific scenarios can effectively optimize vehicle scheduling in advance, significantly shortening pick-up and drop-off times during peak hours and in rainy or snowy weather. This not only improves users' daily travel experience but also reduces platform operating costs and increases the profitability of companies.

1.3 Research Landscape and Open Challenges

This section provides a concise overview of the evolution of spatio-temporal prediction research and the key challenges that have shaped the field. It first reviews classical mechanism-based approaches that emphasize physical interpretability, followed by modern deep learning methods that prioritize predictive accuracy. Building on these perspectives, the section then discusses emerging hybrid modeling paradigms that integrate domain mechanisms into deep learning frameworks, highlighting their potential to address limitations in interpretability, robustness, and generalization.

1.3.1 Mechanism-Based Approaches to Spatio-Temporal Prediction

Early spatio-temporal prediction studies focused on modeling the interaction relationships between regions in geographic space. Spatial interaction is defined as the dynamic flow process of individuals, materials, and information between geographic nodes [45]. Such flow phenomena naturally form a directed flow network embedded in geographic space, i.e. the spatial interaction network [10]. For instance, daily commuting activities form short-term human mobility networks [25], while inter-regional population migration constructs macro-scale population migration networks [9].

Traditional spatio-temporal prediction methods are mainly based on the mechanistic modeling paradigm. They construct parameterized mathematical expressions by analyzing the internal operating mechanisms of the system, and require adaptive adjustments of hypothetical conditions for specific application scenarios. As early as the late 19th century, researchers discovered the universal law of the gravity model in the spatial interaction network of human mobility [4, 80], and further developed the classic gravity model [45, 49]. Different from the gravity model that describes spatial interaction from a macro-statistical perspective, the intervening opportunity model [50] is based on the micro-decision-making mechanism of individuals and constructs an interaction intensity prediction framework based on the spatial search process. Furthermore, to alleviate the dependence of traditional models on empirical data in parameter calibration, researchers abstracted human travel behavior as an energy radiation-absorption process and proposed the radiation model [41, 46].

Although regional-level spatial interaction mechanistic models have important theoretical value, they are difficult to depict road network-level details in real urban traffic. For this reason, research has gradually focused on more practical road network-level traffic flow prediction, and constructed multi-level theoretical models around vehicle behavior and traffic flow evolution processes [11, 36]. At the micro level, typical representatives include car-following models [39, 42] and their extended versions considering factors such as driver memory and multi-vehicle following [28, 35, 51, 68]; at the macro level, they are mainly based on kinematic wave theory, such as the Lighthill-Whitham-Richards (LWR) model [31, 43] and its subsequent improved models, including the Payne model [38] and the Ross model [44].

Literature Review. Although spatio-temporal prediction methods based on mechanistic models have clear physical meanings and can produce interpretable prediction results, they have the following limitations.

Firstly, mechanistic models mainly model simple and static scenarios with ideal assumptions, lacking the ability to depict complex, dynamic processes affected by various factors. However, spatio-temporal data are highly dynamic and nonlinear, which leads to poor spatio-temporal prediction performance of such models.

Secondly, the establishment of mechanistic models requires a great deal of domain knowledge and customized design for specific problems. Nevertheless, some real-world systems are often very complex, and the laws mastered by humans are limited, making it difficult to meet the knowledge reserve requirements for establishing mechanistic models.

1.3.2 Deep Learning Approaches to Spatio-Temporal Prediction

With the continuous development of deep learning technology, its strong ability to automatically learn various data patterns from a large number of samples has attracted the attention of researchers in the field of spatio-temporal prediction. Therefore, an increasing number of deep learning-based spatio-temporal prediction methods have emerged. Since their development, spatio-temporal prediction methods based on deep learning have roughly gone through three stages.

Stage I: Diversified Development. The research work in the first stage usually used various multilayer neural networks to directly perform brute-force prediction on spatio-temporal data. For example, networks based on Deep Belief Networks (DBN) and their improved versions [33, 48, 62], networks based on autoencoders [6], networks using fully connected networks [53], networks adopting Long-Short Term Memory (LSTM) networks [24, 34, 47], and networks utilizing CNN [55, 74].

During this stage, a large number of innovative research works emerged in the field of spatio-temporal prediction. Researchers conducted extensive and in-depth explorations on various feasible deep learning models, accumulating valuable theoretical and practical experience for the development of deep learning models in this field.

Stage II: Paradigm Consolidation. In the second stage, researchers reached a basic consensus that spatio-temporal data should be modeled separately from the temporal and spatial dimensions, i.e., selecting deep learning models specifically to model their temporal dependence and spatial dependence, respectively.

For the modeling of temporal dependence, there are mainly two branches: manual multi-scale modeling and deep automatic modeling. Researchers in the first branch artificially construct temporal dependencies of different scales through data analysis [14, 73, 76], including short-term proximity, medium-term periodicity, and long-term trend. The other branch mainly uses deep neural networks to automatically learn temporal dependencies in data, adopting methods such as RNN and its variants [8, 30, 77] or Temporal Convolutional Networks (TCN) based on 1D CNN [14, 64, 65, 71].

On the other hand, the modeling of spatial dependence includes grid structure-based modeling and graph structure-based modeling. Early researchers simply divided urban areas into grid structures and then used CNN to capture spatial proximity between regions [5, 15, 73]. However, many spatio-temporal data have non-Euclidean spatial structures, such as urban traffic networks (including road networks and subway networks) and meteorological monitoring station networks [17], which are more similar to the topological structure of graphs. This structure is naturally

suitable for Graph Neural Networks GNNs, so there have been many subsequent research works on modeling spatial dependence based on GNNs [12, 13, 30, 71].
Stage III: Innovative Exploration. After approximately five years of development and convergence, researchers found that spatio-temporal prediction seemed to reach a performance bottleneck. Thus, they began to focus on the inherent characteristics of data itself and explore emerging perspectives for spatio-temporal prediction, which indicates that the research in this field has entered the third stage.

There are many cutting-edge perspectives in this stage, among which the most concerned are dynamically changing spatio-temporal dependence [65, 69] and spatio-temporal heterogeneity that considers differences across different spatial regions and time periods [3, 37, 70, 72].
Literature Review. Through the continuous efforts of researchers in the field, deep learning models have achieved notable performance on benchmark datasets in the field of spatio-temporal prediction. However, these methods still have the following limitations.

Firstly, spatio-temporal prediction scenarios usually require prediction models to not only have excellent prediction performance but also elucidate the underlying development and evolution mechanisms hidden in spatio-temporal data. Nevertheless, deep learning models mainly focus on minimizing the error between model predictions and real data, and the issue of internal mechanisms is not within the scope of model design considerations.

Secondly, the design paradigm of existing deep learning prediction models is to use complex "black-box" models to approximate the observed spatio-temporal samples of the real world, i.e., training data, in the hope that the models can generalize to test data. However, considering that the generation process of spatio-temporal data is usually affected by various external scenario factors, this leads to a shift in the distribution of spatio-temporal data between the training data and test data. Consequently, the models exhibit insufficient generalization performance in such scenarios, which greatly limits their applications in real-world settings.

1.3.3 Deep Learning Modeling with Mechanism Integration

Mechanistic models are constructed based on domain knowledge and the fundamental principles of system operation, with their core characteristics being clear physical meanings and strong interpretability. In contrast, deep learning conducts iterative training through a large amount of observational data to mine potential complex relationships in the data, demonstrating excellent nonlinear fitting capabilities without the need for preset assumptions. To fully leverage the advantages of both, researchers have begun to explore modeling strategies that integrate mechanistic models with deep learning, thereby ensuring strong data generalization while taking into account mechanism interpretability. Representative studies in this field include Physics-Informed Neural Networks (PINNs) and Neural Operators (NOs).

Specifically, the core idea of PINN is to introduce physical equations (such as conservation laws and Navier-Stokes equations) as regularizers into the loss function design of neural networks to constrain the output values of the neural network model, ensuring that they satisfy certain physical laws [40]. Since its proposal, PINN has been widely applied to physical modeling and simulation tasks such as partial differential equation solving [23, 78], fluid dynamics simulation [32], and radio map prediction [21, 29].

On the other hand, NOs are essentially derived from operator learning. Operator learning is defined as: given a set of function pairs as data, where one function belongs to the input space and the other belongs to the output space, the goal is to find an approximate representation of the unknown operator. This approximate representation should ensure that for any new input data, a result similar to the real output function can be predicted, with specific examples including derivative operators and Laplacian operators. From the perspective of mechanistic models, the fitted operator is often the core mathematical equation describing the correlation between system input and output; however, in more complex application scenarios, the mathematical equations corresponding to the system are mostly difficult to obtain explicitly. By using neural networks to fit the operator, the limitation of traditional mechanistic models in describing unknown equations is cleverly broken through, providing an effective alternative path for complex system modeling.

Literature Review. The data modeled by PINNs and NOs are usually spatio-temporal data (generated by physical systems), so these methods are naturally applicable to spatio-temporal data prediction tasks. However, spatio-temporal data in urban scenarios are mainly generated by human socioeconomic activities, and they are affected by human behavioral factors and data distribution factors caused by the external environment. The mechanisms behind these factors are rarely considered in existing deep learning methods that integrate mechanistic models.

In summary, in urban spatio-temporal prediction tasks, how to integrate the mechanisms of internal behavioral factors and external scenario factors into the modeling process of deep learning is a major challenge in current research in this field, which requires in-depth exploration.

1.4 Scope and Objectives of the Book

To further improve the generalization performance of deep learning-based spatio-temporal prediction models in urban traffic scenarios and make up for their deficiencies in mechanistic modeling capabilities, this book aims to construct a mechanism-driven deep spatio-temporal prediction paradigm.

Centered on the goal, this book investigates the mechanism factors from two complementary perspectives: intrinsic behavioral mechanisms and extrinsic contextual/scenario dynamics. From these two perspectives, the mechanisms of urban human activities and scenario environment evolution are integrated into the modeling process of deep learning. Thus, an interpretable prediction model with high prediction accuracy that adheres to the inherent mechanisms of data is realized, and the model's generalization ability in complex urban traffic scenarios is enhanced.

This book's research objectives include the following four aspects.

1. To address the issue that existing deep learning models for regional traffic flow prediction have not fully considered the inherent mechanisms of human mobility, this book aims to develop a spatio-temporal potential field extraction model based on gravity theory. This method can extract discrete spatio-temporal potential fields that drive human mobility from urban regional traffic flow according to the inherent mechanism of human mobility—the gravity model. It provides spatio-temporal features containing the mechanisms of human mobility for subsequent deep spatio-temporal prediction models, thereby ensuring the model's prediction performance while endowing it with the ability to provide mechanism-consistent explanations for prediction results.
2. To address the inconsistency between the discrete time assumption and the energy continuity principle in existing road network traffic flow prediction models, this book aims to develop a spatio-temporal differential network based on the continuity equation. Constrained by the energy continuity principle followed by the potential field, this network can embed temporal continuity into the model structure, enabling it to follow the spatio-temporal evolution trajectory of the potential field on the road network over continuous time during the prediction process, thereby improving the accuracy of the model in road network traffic flow prediction.
3. To address the issue that existing traffic state prediction models fail to consider spatio-temporal data distribution shifts caused by confounding factors, this book aims to develop a spatio-temporal basis vector representation model based on causal modeling. Guided by the idea of causal modeling, this model can characterize any unknown confounding factors in the form of basis vectors, thereby separating the impact of confounding factors during the spatio-temporal prediction process and accurately modeling their effects on future data distributions to enhance the robustness and generalization ability of predictions.
4. To address the issue that mixed spatio-temporal distributions formed by the interaction of multiple factors restrict the performance upper limit of existing traffic state prediction models, this book aims to develop a spatio-temporal decomposition prediction framework based on information theory. This framework can use information-theoretic tools to derive the sufficient conditions that an effective decomposition algorithm for reducing data complexity must satisfy. In the decomposition prediction model, it constrains the decomposition process through regularizers to reduce the complexity of input spatio-temporal data, thereby improving the model's prediction performance in complex traffic scenarios and breaking through the performance bottleneck of existing spatio-temporal prediction models.

The research framework and logical relationship of each part of this book are shown in Fig. 1.2.

1.5 Structure of the Book

As shown in Fig. 1.2, this book is divided into seven chapters, which systematically present the research background, methodologies, and key findings.

This chapter serves as the Introduction. It first elaborates on the research background and significance, then traces the development of spatio-temporal prediction methods and deep learning modeling techniques integrated with mechanistic models. Representative research outcomes are reviewed, summarized, and synthesized, followed by an analysis of the limitations of existing methods. This chapter clarifies the research scopes and objectives, highlighting the necessity of the present study. Additionally, based on the research framework, it details the research content and methodologies employed.

Chapter 2 is the Literature Review. It comprehensively reviews domestic and international research progress and overall development in fields relevant to this study, covering spatio-temporal prediction methods, deep learning integrated with mechanistic models, spatio-temporal distribution shift, and spatio-temporal decomposition prediction.

Chapter 3 focuses on the spatio-temporal potential field extraction model based on gravity theory. Firstly, it proposes the concept of a spatio-temporal potential

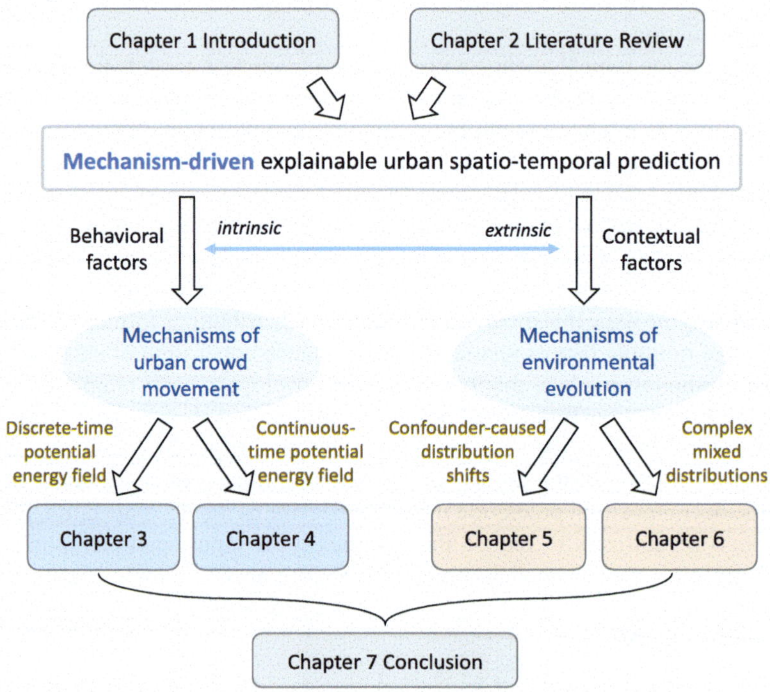

Fig. 1.2 Research framework and structure of this book

field rooted in gravity theory and explains its core ideology. Subsequently, it details the discrete-time regional potential field extraction algorithm and the correlation-adaptive spatio-temporal modeling method. Finally, extensive experiments verify that the spatio-temporal potential field can effectively characterize the inherent mechanisms of human mobility and enhance the performance of discrete-time regional traffic flow prediction.

Chapter 4 investigates the spatio-temporal differential network model based on the continuity equation. Initially, it derives the differential equation governing the road network potential field from the energy continuity equation. Then, guided by this potential field differential equation, it designs a spatio-temporal differential network with continuous hidden layers, elaborating on the network structure, parameter learning, and prediction process. Experiments ultimately confirm that the proposed model can effectively capture the potential evolution of road network traffic flow and improve generalization on continuous-time data.

Chapter 5 studies the spatio-temporal basis vector representation model based on causal modeling. It first introduces a new definition of the traffic prediction problem considering the influence of confounders. Next, it designs a confounder extractor that leverages the basis vector ideology and self-supervised learning to acquire representations of arbitrary confounders, while an adversarial decoupling module captures confounder-agnostic spatio-temporal relationships. Experimental results demonstrate the model's robust prediction performance in scenarios with spatio-temporal distribution shifts induced by confounders.

Chapter 6 explores the spatio-temporal decomposition prediction framework based on information theory. It begins by introducing the traffic prediction problem from an information-theoretic perspective, then proposes the decomposition component independence principle based on Fano's inequality through theoretical analysis. Further, it develops a generalizable spatio-temporal graph decomposition learning framework grounded in this principle. Experiments verify the framework's universality across different spatio-temporal prediction models and its effectiveness in modeling mixed distributions formed by multiple influencing factors.

The final chapter summarizes the core content of the book, synthesizes its innovations, research findings, and contributions from each chapter. It also identifies the study's limitations and puts forward future research directions and suggestions, providing valuable insights and references for subsequent studies.

References

1. Alam, M.M., Torgo, L., Bifet, A.: A survey on spatio-temporal data analytics systems. ACM Computing Surveys **54**(10s), 1–38 (2022)
2. Atluri, G., Karpatne, A., Kumar, V.: Spatio-temporal data mining: A survey of problems and methods. ACM Computing Surveys **51**(4), 1–41 (2018)
3. Bai, L., Yao, L., Li, C., Wang, X., Wang, C.: Adaptive graph convolutional recurrent network for traffic forecasting. In: Proceedings of the 34th International Conference on Neural Information Processing Systems (NeurIPS'20), pp. 17804–17815 (2020)

4. Carey, H.C.: Principles of social science. JB Lippincott and Company (1871)
5. Chen, C., Li, K., Teo, S.G., Chen, G., Zou, X., Yang, X., Vijay, R.C., Feng, J., Zeng, Z.: Exploiting spatio-temporal correlations with multiple 3D convolutional neural networks for citywide vehicle flow prediction. In: Proceedings of the 2018 IEEE International Conference on Data Mining (ICDM'18), pp. 893–898 (2018)
6. Chen, Q., Song, X., Yamada, H., Shibasaki, R.: Learning deep representation from big and heterogeneous data for traffic accident inference. In: Proceedings of the 30th AAAI Conference on Artificial Intelligence (AAAI'16), pp. 338–344 (2016)
7. Cho, K., van Merriënboer, B., Gulcehre, C., Bahdanau, D., Bougares, F., Schwenk, H., Bengio, Y.: Learning phrase representations using RNN encoder–decoder for statistical machine translation. In: Proceedings of the 2014 Conference on Empirical Methods in Natural Language Processing, pp. 1724–1734 (2014)
8. Cui, Z., Henrickson, K., Ke, R., Wang, Y.: Traffic graph convolutional recurrent neural network: A deep learning framework for network-scale traffic learning and forecasting. IEEE Transactions on Intelligent Transportation Systems (ITS) **21**(11), 4883–4894 (2019)
9. Davis, K.F., D'Odorico, P., Laio, F., Ridolfi, L.: Global spatio-temporal patterns in human migration: A complex network perspective. PLOS One **8**(1), e53723 (2013)
10. Dejon, B.: Spatial interaction network flow models. In: Proceedings of the Annual Meeting 1977 DGOR, pp. 377–386 (1978)
11. Elefteriadou, L., et al.: An introduction to traffic flow theory. Springer (2014)
12. Geng, X., Li, Y., Wang, L., Zhang, L., Yang, Q., Ye, J., Liu, Y.: Spatiotemporal multi-graph convolution network for ride-hailing demand forecasting. In: Proceedings of the 33rd AAAI Conference on Artificial Intelligence (AAAI'19), pp. 3656–3663 (2019)
13. Guo, K., Hu, Y., Sun, Y., Qian, S., Gao, J., Yin, B.: Hierarchical graph convolution network for traffic forecasting. In: Proceedings of the 35th AAAI Conference on Artificial Intelligence (AAAI'21), pp. 151–159 (2021)
14. Guo, S., Lin, Y., Feng, N., Song, C., Wan, H.: Attention based spatial-temporal graph convolutional networks for traffic flow forecasting. In: Proceedings of the 33rd AAAI Conference on Artificial Intelligence (AAAI'19), pp. 922–929 (2019)
15. Guo, S., Lin, Y., Li, S., Chen, Z., Wan, H.: Deep spatial-temporal 3D convolutional neural networks for traffic data forecasting. IEEE Transactions on Intelligent Transportation Systems (ITS) **20**(10), 3913–3926 (2019)
16. He, K., Zhang, X., Ren, S., Sun, J.: Deep residual learning for image recognition. In: Proceedings of the 2016 IEEE Conference on Computer Vision and Pattern Recognition (CVPR'16), pp. 770–778 (2016)
17. Hettige, K.H., Ji, J., Xiang, S., Long, C., Cong, G., Wang, J.: AirPhyNet: Harnessing physics-guided neural networks for air quality prediction. In: Proceedings of the 12th International Conference on Learning Representations (ICLR'24), pp. 1–17 (2024)
18. Hochreiter, S., Schmidhuber, J.: Long short-term memory. Neural Computation **9**(8), 1735–1780 (1997)
19. Jain, L.C., Medsker, L.R.: Recurrent neural networks: Design and applications. CRC Press (1999)
20. Ji, J., Wang, J., Wu, J., Han, B., Zhang, J., Zheng, Y.: Precision CityShield against hazardous chemicals threats via location mining and self-supervised learning. In: Proceedings of the 28th ACM SIGKDD Conference on Knowledge Discovery and Data Mining (KDD'22), pp. 3072–3080 (2022)
21. Jiang, F., Li, T., Lv, X., Rui, H., Jin, D.: Physics-informed neural networks for path loss estimation by solving electromagnetic integral equations. IEEE Transactions on Wireless Communications (TWC) **23**(1), 15380–15393 (2024)
22. Jin, G., Liang, Y., Fang, Y., Shao, Z., Huang, J., Zhang, J., Zheng, Y.: Spatio-temporal graph neural networks for predictive learning in urban computing: A survey. IEEE Transactions on Knowledge and Data Engineering (TKDE) **36**(10), 5388–5408 (2024)
23. Kharazmi, E., Zhang, Z., Karniadakis, G.E.: hp-VPINNs: Variational physics-informed neural networks with domain decomposition. Computer Methods in Applied Mechanics and Engineering **374**(1), 113547 (2021)

24. Kong, D., Wu, F.: HST-LSTM: A hierarchical spatial-temporal long-short term memory network for location prediction. In: Proceedings of the 27th International Joint Conference on Artificial Intelligence (IJCAI'18), pp. 2341–2347 (2018)
25. Kwan, M.P.: Mobile communications, social networks, and urban travel: Hypertext as a new metaphor for conceptualizing spatial interaction. The Professional Geographer **59**(4), 434–446 (2007)
26. LeCun, Y., Bengio, Y., Hinton, G.: Deep learning. Nature **521**(7553), 436–444 (2015)
27. LeCun, Y., Boser, B., Denker, J., Henderson, D., Howard, R., Hubbard, W., Jackel, L.: Handwritten digit recognition with a back-propagation network. In: Proceedings of the 3rd International Conference on Neural Information Processing Systems (NeurIPS'89), p. 396–404 (1989)
28. Lee, G.: A generalization of linear car-following theory. Operations Research **14**(4), 595–606 (1966)
29. Levie, R., Yapar, Ç., Kutyniok, G., Caire, G.: RadioUNet: Fast radio map estimation with convolutional neural networks. IEEE Transactions on Wireless Communications (TWC) **20**(6), 4001–4015 (2021)
30. Li, Y., Yu, R., Shahabi, C., Liu, Y.: Diffusion convolutional recurrent neural network: Data-driven traffic forecasting. In: Proceedings of the 6th International Conference on Learning Representations (ICLR'18), pp. 1–16 (2018)
31. Lighthill, M.J., Whitham, G.B.: On kinematic waves II. A theory of traffic flow on long crowded roads. Proceedings of the Royal Society of London **229**(1178), 317–345 (1955)
32. Lu, L., Jin, P., Pang, G., Zhang, Z., Karniadakis, G.E.: Learning nonlinear operators via DeepONet based on the universal approximation theorem of operators. Nature Machine Intelligence **3**(3), 218–229 (2021)
33. Lv, Y., Duan, Y., Kang, W., Li, Z., Wang, F.Y.: Traffic flow prediction with big data: A deep learning approach. IEEE Transactions on Intelligent Transportation Systems (ITS) **16**(2), 865–873 (2014)
34. Ma, X., Tao, Z., Wang, Y., Yu, H., Wang, Y.: Long short-term memory neural network for traffic speed prediction using remote microwave sensor data. Transportation Research Part C: Emerging Technologies **54**(1), 187–197 (2015)
35. Newell, G.F.: Nonlinear effects in the dynamics of car following. Operations Research **9**(2), 209–229 (1961)
36. Ni, D.: Traffic flow theory: Characteristics, experimental methods, and numerical techniques. Butterworth-Heinemann (2015)
37. Pan, Z., Wang, Z., Wang, W., Yu, Y., Zhang, J., Zheng, Y.: Matrix factorization for spatio-temporal neural networks with applications to urban flow prediction. In: Proceedings of the 28th ACM International Conference on Information and Knowledge Management (CIKM'19), pp. 2683–2691 (2019)
38. Payne, H.J.: Freeway traffic control and surveillance model. Transportation Engineering Journal of ASCE **99**(4), 767–783 (1973)
39. Pipes, L.A.: An operational analysis of traffic dynamics. Journal of Applied Physics **24**(3), 274–281 (1953)
40. Raissi, M., Perdikaris, P., Karniadakis, G.E.: Physics-informed neural networks: A deep learning framework for solving forward and inverse problems involving nonlinear partial differential equations. Journal of Computational Physics **378**(1), 686–707 (2019)
41. Ren, Y., Ercsey-Ravasz, M., Wang, P., González, M.C., Toroczkai, Z.: Predicting commuter flows in spatial networks using a radiation model based on temporal ranges. Nature Communications **5**(1), 5347 (2014)
42. Reuschel, A.: Vehicle movements in the column uniformly accelerated or delayed. Oesterrich Ingr Arch **4**(1), 193–215 (1950)
43. Richards, P.I.: Shock waves on the highway. Operations Research **4**(1), 42–51 (1956)
44. Ross, P.: Traffic dynamics. Transportation Research Part B: Methodological **22**(6), 421–435 (1988)
45. Roy, J.R., Thill, J.C.: Spatial Interaction Modelling. Springer (2004)

46. Simini, F., González, M.C., Maritan, A., Barabási, A.L.: A universal model for mobility and migration patterns. Nature **484**(7392), 96–100 (2012)
47. Song, X., Kanasugi, H., Shibasaki, R.: DeepTransport: Prediction and simulation of human mobility and transportation mode at a citywide level. In: Proceedings of the 25th International Joint Conference on Artificial Intelligence (IJCAI'16), pp. 2618–2624 (2016)
48. Soua, R., Koesdwiady, A., Karray, F.: Big-data-generated traffic flow prediction using deep learning and Dempster-Shafer theory. In: Proceedings of the 2016 International Joint Conference on Neural Networks, pp. 3195–3202 (2016)
49. Stewart, J.Q.: An inverse distance variation for certain social influences. Science **93**(2404), 89–90 (1941)
50. Stouffer, S.A.: Intervening opportunities: A theory relating mobility and distance. American Sociological Review **5**(6), 845–867 (1940)
51. Toledo, T., Koutsopoulos, H.N., Ben-Akiva, M.: Integrated driving behavior modeling. Transportation Research Part C: Emerging Technologies **15**(2), 96–112 (2007)
52. Vaswani, A., Shazeer, N., Parmar, N., Uszkoreit, J., Jones, L., Gomez, A.N., Kaiser, L., Polosukhin, I.: Attention is all you need. In: Proceedings of the 31st International Conference on Neural Information Processing Systems (NeurIPS'17), pp. 5998–6008 (2017)
53. Wang, D., Cao, W., Li, J., Ye, J.: DeepSD: Supply-demand prediction for online car-hailing services using deep neural networks. In: Proceedings of the 2017 IEEE 33rd International Conference on Data Engineering (ICDE'17), pp. 243–254 (2017)
54. Wang, J., Chen, C., Wu, J., Xiong, Z.: No longer sleeping with a bomb: A duet system for protecting urban safety from dangerous goods. In: Proceedings of the 23rd ACM SIGKDD Conference on Knowledge Discovery and Data Mining (KDD'17), pp. 1673–1681 (2017)
55. Wang, J., Gu, Q., Wu, J., Liu, G., Xiong, Z.: Traffic speed prediction and congestion source exploration: A deep learning method. In: Proceedings of the 2016 IEEE International Conference on Data Mining (ICDM'16), pp. 499–508 (2016)
56. Wang, J., Lin, X., Zuo, Y., Wu, J.: DGeye: Probabilistic risk perception and prediction for urban dangerous goods management. ACM Transactions on Information Systems (TOIS) **39**(3), 1–30 (2021)
57. Wang, J., Shi, H., Ji, J., Lin, X., Tian, H.: High-resolution data on human behavior for effective COVID-19 policy-making—Wuhan city, Hubei province, China, January 1–February 29, 2020. China CDC Weekly **5**(4), 76–81 (2023)
58. Wang, J., Tang, K., Feng, K., Lin, X., Lv, W., Chen, K., Wang, F.: Impact of temperature and relative humidity on the transmission of COVID-19: A modelling study in China and the United States. BMJ Open **11**(2), e043863 (2021)
59. Wang, J., Wang, X., Wu, J.: Inferring metapopulation propagation network for intra-city epidemic control and prevention. In: Proceedings of the 24th ACM SIGKDD Conference on Knowledge Discovery and Data Mining (KDD'18), pp. 830–838 (2018)
60. Wang, S., Cao, J., Philip, S.Y.: Deep learning for spatio-temporal data mining: A survey. IEEE Transactions on Knowledge and Data Engineering (TKDE) **34**(8), 3681–3700 (2020)
61. Wang, Z., Jiang, R., Xue, H., Salim, F.D., Song, X., Shibasaki, R.: Event-aware multimodal mobility nowcasting. In: Proceedings of the 36th AAAI Conference on Artificial Intelligence (AAAI'22), pp. 4228–4236 (2022)
62. Wei, H., Wang, Y., Wo, T., Liu, Y., Xu, J.: ZEST: A hybrid model on predicting passenger demand for chauffeured car service. In: Proceedings of the 25th ACM International Conference on Information and Knowledge Management (CIKM'16), pp. 2203–2208 (2016)
63. Wu, Z., Pan, S., Chen, F., Long, G., Zhang, C., Philip, S.Y.: A comprehensive survey on graph neural networks. IEEE Transactions on Neural Networks and Learning Systems **32**(1), 4–24 (2020)
64. Wu, Z., Pan, S., Long, G., Jiang, J., Chang, X., Zhang, C.: Connecting the dots: Multivariate time series forecasting with graph neural networks. In: Proceedings of the 26th ACM SIGKDD Conference on Knowledge Discovery and Data Mining (KDD'20), pp. 753–763 (2020)
65. Wu, Z., Pan, S., Long, G., Jiang, J., Zhang, C.: Graph WaveNet for deep spatial-temporal graph modeling. In: Proceedings of the 28th International Joint Conference on Artificial Intelligence (IJCAI'19), pp. 1907–1913 (2019)

66. Wu, Z., Wu, L., Song, S., Ji, J., Zou, B., Li, Z., He, X.: DialCSP: A two-stage attention-based model for customer satisfaction prediction in e-commerce customer service. In: Proceedings of the 2022 Joint European Conference on Machine Learning and Knowledge Discovery in Databases, pp. 3–18 (2022)
67. Wu, Z., Yu, X., Chen, M., Wu, L., Ji, J., Li, Z.: Enhancing new intent discovery via robust neighbor-based contrastive learning. In: Proceedings of the 24th Annual Conference of the International Speech Communication Association, pp. 740–744 (2023)
68. Yang, Q., Koutsopoulos, H.N.: A microscopic traffic simulator for evaluation of dynamic traffic management systems. Transportation Research Part C: Emerging Technologies **4**(3), 113–129 (1996)
69. Yao, H., Tang, X., Wei, H., Zheng, G., Li, Z.: Revisiting spatial-temporal similarity: A deep learning framework for traffic prediction. In: Proceedings of the 33rd AAAI Conference on Artificial Intelligence (AAAI'19), pp. 5668–5675 (2019)
70. Ye, X., Fang, S., Sun, F., Zhang, C., Xiang, S.: Meta graph transformer: A novel framework for spatial-temporal traffic prediction. Neurocomputing **491**(1), 544–563 (2022)
71. Yu, B., Yin, H., Zhu, Z.: Spatio-temporal graph convolutional networks: A deep learning framework for traffic forecasting. In: Proceedings of the 27th International Joint Conference on Artificial Intelligence (IJCAI'18), pp. 3634–3640 (2018)
72. Yuan, Z., Zhou, X., Yang, T.: Hetero-ConvLSTM: A deep learning approach to traffic accident prediction on heterogeneous spatio-temporal data. In: Proceedings of the 24th ACM SIGKDD Conference on Knowledge Discovery and Data Mining (KDD'18), pp. 984–992 (2018)
73. Zhang, J., Zheng, Y., Qi, D.: Deep spatio-temporal residual networks for citywide crowd flows prediction. In: Proceedings of the 31st AAAI Conference on Artificial Intelligence (AAAI'17), pp. 1655–1661 (2017)
74. Zhang, J., Zheng, Y., Qi, D., Li, R., Yi, X.: DNN-based prediction model for spatio-temporal data. In: Proceedings of the 24th ACM SIGSPATIAL International Conference on Advances in Geographic Information Systems, pp. 1–4 (2016)
75. Zhang, X., Cui, P., Xu, R., Zhou, L., He, Y., Shen, Z.: Deep stable learning for out-of-distribution generalization. In: Proceedings of the 2021 IEEE Conference on Computer Vision and Pattern Recognition (CVPR'21), pp. 5372–5382 (2021)
76. Zhang, X., Huang, C., Xu, Y., Xia, L., Dai, P., Bo, L., Zhang, J., Zheng, Y.: Traffic flow forecasting with spatial-temporal graph diffusion network. In: Proceedings of the 35th AAAI Conference on Artificial Intelligence (AAAI'21), pp. 15008–15015 (2021)
77. Zhao, L., Song, Y., Zhang, C., Liu, Y., Wang, P., Lin, T., Deng, M., Li, H.: T-GCN: A temporal graph convolutional network for traffic prediction. IEEE Transactions on Intelligent Transportation Systems (ITS) **21**(9), 3848–3858 (2019)
78. Zhao, Z., Ding, X., Prakash, B.A.: PINNsFormer: A transformer-based framework for physics-informed neural networks. In: Proceedings of the 12th International Conference on Learning Representations (ICLR'24), pp. 1–17 (2024)
79. Zheng, Y.: Urban computing. MIT Press (2019)
80. Zipf, G.K.: The $P_1 P_2/D$ hypothesis: On the intercity movement of persons. American Sociological Review **11**(6), 677–686 (1946)

Chapter 2
Literature Review of Mechanism-Driven and Learning-Driven Prediction

Abstract This chapter first reviews the development of spatiotemporal prediction methods, mechanism-aware deep learning models, spatiotemporal distribution-shift modeling techniques, and spatiotemporal decomposition-based forecasting approaches. Building on this survey, it then summarizes the limitations of representative methods along each of these lines of work.

2.1 Overview of Spatio-Temporal Prediction Methods

Spatio-temporal prediction is a key technique for enabling scientific, fine-grained, and intelligent urban management, and is particularly central to transportation planning and control. In recent years, with the rapid accumulation of spatio-temporal data in the transportation domain, prediction methods targeting traffic data have attracted considerable attention [44].

According to the type of underlying model, existing spatio-temporal prediction approaches can be broadly divided into interpretable, mechanism-based models and data-driven deep learning models. This section reviews these two lines of work, tracing their historical development and identifying their limitations so as to reveal opportunities for further innovation.

2.1.1 Mechanism-Based Methods

Research on spatio-temporal prediction can be traced back to early studies of spatial interaction between geographic regions [92]. These classical models typically operate at a coarse spatial granularity and are mainly used for macroscopic tasks such as urban planning and public transit scheduling. As the demand for fine-grained urban management has grown, the focus of spatio-temporal prediction has gradually shifted toward finer spatial scales, with road network-level traffic flow prediction becoming one of the most representative tasks.

J. Wang and J. Ji, *Mechanism-Driven Explainable Urban Spatio-Temporal Prediction*, Big Data Management, https://doi.org/10.1007/978-981-92-0662-9_2

Accordingly, this section first reviews mechanism-based models for coarse-grained, region-level spatial interaction prediction, and then turns to modern traffic flow theory [70] and mechanism-based models for fine-grained, road network-level traffic flow prediction.

2.1.1.1 Region-Level Spatial Interaction Prediction

Region-level spatial interaction refers to the dynamic flow of individuals, goods, and information between different geographic locations [92]. Such flows naturally form a directed network embedded in geographic space, commonly termed a spatial interaction network [21]. For example, daily commuting activities induce short-term crowd mobility networks [50], whereas inter-city or international migration gives rise to large-scale population migration networks [19]. The evolution of these spatial interaction networks not only reveals the intrinsic regularities of human social activities, but also provides important guidance for transportation planning, resource allocation, and policy-making.

Early work on spatio-temporal prediction predominantly relied on mechanism-based models for analysis and forecasting. These models formalize the underlying laws of human mobility in mathematical terms and provide theoretical support for prediction and optimization. Representative mechanism-based models include the classical gravity model, the intervening opportunities model that emphasizes individual decision processes, and the more recent radiation model. This subsection provides a systematic review of these three model families, covering their mathematical formulations, theoretical foundations, and application characteristics.

Gravity Model

The gravity model [92, 102] is one of the most fundamental theories for spatial interaction modeling, whose mathematical form directly parallels Newton's law of universal gravitation. Its core formulation is

$$T_{ij} = \alpha \frac{m_i m_j}{d_{ij}^{\beta}}, \tag{2.1}$$

where T_{ij} denotes the flow from origin i to destination j, m_i and m_j denote the population size or economic indicators of the origin and destination, respectively, d_{ij} is the spatial distance or generalized travel cost between the two locations, α is a global scaling factor, and β is a distance decay exponent [126]. The core hypothesis can be summarized as follows: the interaction intensity between two regions in terms of population movement, travel, or trade is positively correlated with their population sizes and negatively correlated with the β-th power of the geographical distance.

The original gravity model in Eq. (2.1) cannot guarantee that the resulting flow matrix $\mathbf{T}$ satisfies both the origin outflow and destination inflow constraints. To address this issue, subsequent work introduced balancing factors A_i and B_j into the classical model, thereby preserving the core mechanism of the gravity model while ensuring flow conservation at origins and destinations [92]. The modified gravity model is written as

$$T_{ij} = A_i B_j O_i D_j f(d_{ij}), \tag{2.2}$$

where $f(d_{ij})$ denotes the generalized travel cost, commonly instantiated as $d_{ij}^{-\beta}$, $e^{-\beta d_{ij}}$, and related forms [95]. Here $O_i = \sum_j T_{ij}$ is the total outflow from origin i and $D_j = \sum_i T_{ij}$ is the total inflow to destination j. The balancing factors are given by

$$A_i = \left[\sum_j B_j D_j f(d_{ij}) \right]^{-1}, \quad B_j = \left[\sum_i A_i O_i f(d_{ij}) \right]^{-1}, \tag{2.3}$$

which can be obtained via iterative algorithms. These factors enforce origin–destination flow balance while still accounting for geographic distance and other factors, thereby enhancing the practical value of the model.

Because of its simple form and intuitive interpretation, the gravity model has become the most widely used model for spatial interaction [92]. Although it is constructed by analogy to Newtonian gravitation, the original model lacks a rigorous theoretical justification [97]. To address this, Wilson [119] derived the gravity model using the maximum entropy principle. Under constraints on total travel cost $C = \sum_i \sum_j T_{ij} c_{ij}$, origin totals $O_i = \sum_j T_{ij}$, and destination totals $D_j = \sum_i T_{ij}$, the number of microstates of the trip distribution is

$$\Omega = \frac{T!}{\prod_{ij} T_{ij}!}. \tag{2.4}$$

By solving the constrained optimization problem using the maximum entropy principle [119], one obtains the doubly constrained gravity model

$$T_{ij} = A_i B_j O_i D_j e^{-\beta c_{ij}}, \tag{2.5}$$

where β is a cost-sensitivity parameter. When multiple travel modes are considered, the generalized travel cost is approximately a logarithmic–linear function of distance [127], i.e., $c_{ij} \approx \eta \ln d_{ij} + \mu d_{ij} + \psi$. Substituting this into Eq. (2.5) yields

$$T_{ij} \approx A_i B_j O_i D_j d_{ij}^{\alpha} e^{-\beta d_{ij}}, \tag{2.6}$$

whose form covers most empirical distance decay functions used in gravity models [4] and exhibits good fitting performance in many real-world applications.

Although the maximum entropy principle provides a statistical-physics foundation for the gravity model, it remains essentially a macroscopic statistical description rather than a micro-mechanism model [119]. Specifically, it does not explicitly account for individual-level decision processes, and the physical interpretation of the total cost constraint C is not entirely clear [39]. As a result, the underlying mechanism of individual interaction decisions remains theoretically unsatisfactory [39,

97]. This limitation has motivated subsequent research to focus on individual-level decision processes, leading to the development of micro-mechanism models such as the intervening opportunities model and the radiation model.

Intervening Opportunities Model

The intervening opportunities model [103] reconstructs spatial interaction from a microscopic decision-making perspective. Its core assumption is that travelers scan potential destinations in order of spatial proximity; at each candidate location j, they are attracted to stay with probability αm_j, where α is a stopping probability coefficient and m_j is the population of location j.

Let O_i denote the total number of trips originating from i, and index candidate destinations by increasing distance as $j = 1, 2, \ldots, N$. Denote by q_i^{j-1} the probability that a trip from i has not been attracted by the first $j - 1$ locations. Then the recursion

$$q_i^j = q_i^{j-1}(1 - \alpha m_j) \tag{2.7}$$

holds. Let $S_{i,j}$ be the cumulative population along the route from origin i to destination j, so that $m_j = S_{i,j} - S_{i,j-1}$. We then obtain

$$\frac{q_i^j - q_i^{j-1}}{S_{i,j} - S_{i,j-1}} = -\alpha q_i^{j-1}. \tag{2.8}$$

Treating q and S as continuous variables, this relation can be written as the differential equation

$$\frac{\partial q_i}{\partial S} = -\alpha q_i(S). \tag{2.9}$$

Solving this equation by separation of variables yields an exponential decay form:

$$q_i(S) = \frac{e^{-\alpha S}}{1 - e^{-\alpha M}}, \tag{2.10}$$

where M is the total population in the system. The probability of choosing the j-th destination can thus be written as

$$T_{ij} = O_i \frac{e^{-\alpha(S_{i,j} - m_j)} - e^{-\alpha S_{i,j}}}{1 - e^{-\alpha M}}. \tag{2.11}$$

This model replaces the distance-based decay term $d_{ij}^{-\beta}$ in the classical gravity model (see Eq. (2.1)) with an exponential function of cumulative opportunities $e^{-\alpha S_{i,j}}$, thereby capturing the shielding effect of intervening opportunities between the origin and distant destinations. The core parameter α can be interpreted as the decay strength contributed by a unit population.

Compared with the gravity model, the intervening opportunities model focuses more directly on the sequential destination-choice process of individuals and can be regarded as a prototypical example of micro-mechanism modeling. By ranking candidate destinations according to distance and introducing a recursive stopping process, it explicitly characterizes the decision chain through which individuals evaluate and select opportunities.

However, the intervening opportunities model is mathematically more involved, requiring differential-equation analysis and iterative parameter estimation. Moreover, its strict assumption of sequential spatial scanning often leads to a systematic underestimation of long-distance travel. These limitations have constrained its application in practical transportation planning. Nevertheless, the core idea of replacing geographic distance with the ranking of opportunities as a measure of destination attractiveness has important theoretical implications. By breaking away from the absolute notion of geographic distance, this modeling paradigm directly inspired subsequent parameter-free models such as the radiation model and opened up new methodological perspectives for studying spatial interaction networks.

Radiation Model

In traditional spatial interaction modeling, parameter calibration relies heavily on historical flow data. This limitation motivated the proposal of the parameter-free radiation model [98].

The radiation model assumes that an individual originating at i chooses the nearest destination j whose benefit (proportional to the local population m_j) exceeds that of the origin. Let S_{ij} be the total population between the origin i and destination j. The probability that a trip has not been absorbed by any of the first $j-1$ locations is

$$q_i^j = q_i^{j-1}\left(1 - \frac{m_j}{S_{ij}}\right) = \prod_{k=1}^{j-1} \frac{S_{i,j-1}}{S_{ij}}. \tag{2.12}$$

This leads to the probability of choosing destination j,

$$P_{ij} = q_i^{j-1}\frac{m_j}{S_{ij}} = \frac{m_i m_j}{S_{i,j-1}S_{ij}}. \tag{2.13}$$

The formula links interaction strength to the population distribution: the numerator $m_i m_j$ reflects the scale effect of the origin and destination, while the denominator $S_{i,j-1}S_{ij}$ captures the attenuation effect contributed by the population along the path. This mechanism is analogous to a particle radiation process: travelers are like particles emitted from the origin, and the intermediate population S_{ij} acts as an "absorbing medium" that determines the distribution of stopping locations [73].

Unlike the previous two model families, the radiation model does not require parameter calibration; given only the population set $\{m_j\}$, it can predict spatial interaction flows. Empirical studies on commuting, migration, freight, and other application scenarios show that the radiation model can accurately predict traffic flows across various types of cities [98].

2.1.1.2 Road Network-Level Traffic Flow Prediction

Although mechanism-based models for region-level spatial interaction prediction are of great theoretical importance in spatiotemporal modeling, they face notable limitations in practical applications. In particular, they struggle to capture fine-grained variations at the road-network level in real urban traffic systems. Consequently, recent research has increasingly focused on more realistic road network-level traffic flow prediction, leading to a rich body of traffic flow theories that describe vehicle behavior and the dynamic evolution of traffic flows. These theories provide a solid foundation for traffic system modeling, performance evaluation, and intelligent traffic control [25, 72]. According to the modeling granularity, mechanism-based models for road network-level traffic flow prediction are commonly divided into microscopic and macroscopic models, which respectively capture traffic phenomena at the individual-vehicle level and at the aggregate flow level [36].

Microscopic Model

Microscopic models use individual vehicles as the basic modeling units and focus on the interaction mechanisms between vehicles, especially the dynamics of car-following and lane-changing [107]. Among them, car-following (CF) models are the most representative class of microscopic traffic flow models [8]. They assume that vehicles travel in a fixed order on a single lane where lane-changing is not allowed, and that drivers adjust their behavior in response to the state of the vehicle in front. Car-following models typically use dynamical systems to study how changes in the leading vehicle's (LV) motion state induce behavioral responses in the following vehicle (FV) [87]. By characterizing the step-by-step responses in a platoon, they reveal the microscopic evolution of single-lane traffic flow and bridge individual driving behavior with macroscopic traffic phenomena.

Car-following models primarily describe platoon characteristics in non-free-flow conditions, where the traffic exhibits constrained, delayed, and propagative behavior. As a result, information propagation in a vehicle platoon is not smooth and continuous, but takes the form of wave-like, intermittent pulses [79]. In line with this property, early car-following models commonly adopt a "stimulus–response" paradigm: the influence of the leading vehicle is abstracted as a stimulus s_t, the driver's perception ability is quantified by a sensitivity coefficient s_e, and the driver's response r_e is reflected in the motion state of the following vehicle:

$$r_e = s_t \cdot s_e. \tag{2.14}$$

This classical framework captures the essential mechanism of car-following behavior.

Within this framework, the Gazis–Herman–Rothery (GHR) model is one of the most influential classical models [30]. Its basic formulation is

$$a_n(t+T) = \frac{\lambda v_n^m(t+T)}{[x_{n-1}(t) - x_n(t)]^l} \left[v_{n-1}(t) - v_n(t)\right], \tag{2.15}$$

where $a_n(t+T)$ denotes the acceleration of the n-th vehicle at time $t+T$, and $v_n(t)$ and $x_n(t)$ are its speed and position at time t, respectively. The parameter λ is a sensitivity coefficient, T is the driver's reaction time, and m and l are model parameters estimated from data. Since speed and acceleration are time derivatives of position, Eq. (2.15) can equivalently be written as a differential equation:

$$\frac{\partial^2 x_n(t+T)}{\partial t^2} = \frac{\lambda}{[x_{n-1}(t) - x_n(t)]^l} \left[\frac{\partial x_n(t+T)}{\partial t}\right]^m \left[\frac{\partial x_{n-1}(t)}{\partial t} - \frac{\partial x_n(t+T)}{\partial t}\right]. \tag{2.16}$$

Solving this model yields the acceleration, speed, and relative position of each following vehicle in the platoon at any given time. Based on the spacing and speed difference between a follower and its leader, one can further assess the risk of rear-end collisions and determine whether the traffic stream operates within safe conditions. These theoretical results provide valuable references for determining minimum safe headways and critical reaction times [5, 32, 109].

In the GHR model, the relative speed between the leader and the follower serves as the stimulus term, while the follower's speed and headway jointly determine the sensitivity. A large body of subsequent car-following models extends and refines the GHR framework to incorporate acceleration-deceleration asymmetry [105, 129], driver memory effects [53], multi-vehicle car-following behavior [71], and other behavioral factors. These extensions substantially improve model fidelity and broaden their applicability across diverse traffic scenarios.

Despite substantial theoretical and practical progress, existing car-following models still face important limitations. For instance, they struggle to capture overtaking and lane-changing maneuvers in realistic traffic, and they do not easily integrate road topology (e.g., number of lanes, pavement condition, traffic signals) or driver-specific characteristics such as perception, judgment, and rationality. Moreover, their applicability to complex, heterogeneous traffic environments—especially those prevalent in many developing countries—remains limited. Consequently, car-following models are still best viewed as idealized theoretical frameworks, and their shortcomings also highlight the broader limitations of microscopic traffic flow models.

Macroscopic Model

In contrast to microscopic models that focus on individual driver behavior, macroscopic models emphasize aggregate traffic flow characteristics. They treat traffic streams as continuous media and use partial differential equations to describe the evolution of macroscopic variables such as traffic density and flow over space and time.

This line of research originates from the classical Lighthill-Whitham-Richards (LWR) kinematic wave model proposed by Lighthill, Whitham, and Richards [63, 88]. Lighthill and Whitham first introduced the principle of mass conservation from fluid mechanics into traffic flow modeling, arguing that if there is no inflow or outflow of vehicles between two observation points, the traffic flow must satisfy a conservation law. By analogy, vehicles are treated as a continuous mass, and the number of vehicles per unit road length defines the traffic density. Building on this idea, they derived the continuity equation that describes how traffic states evolve in space and time:

$$\frac{\partial k}{\partial t} + \frac{\partial q}{\partial x} = 0, \tag{2.17}$$

where x and t denote space and time, k is traffic density, and q is traffic flow. Let u denote the average speed along the road; then flow and density are related by

$$q = ku. \tag{2.18}$$

The continuity equation is a dynamic model capable of describing basic traffic phenomena. However, it assumes that vehicle speed is always in equilibrium with density and neglects acceleration and inertia effects. As a result, it cannot adequately capture the dynamic features of non-equilibrium traffic such as stop-and-go waves, and thus has inherent limitations.

To overcome these limitations, researchers have introduced acceleration terms or momentum equations to refine the models. A milestone in this line of work is the general expression for traffic acceleration proposed by Pipes [80]:

$$\frac{\mathrm{d}u}{\mathrm{d}t} = \frac{\partial u}{\partial t} + u\frac{\partial u}{\partial x} = -k\left(\frac{\mathrm{d}u}{\mathrm{d}k}\right)^2 \frac{\partial k}{\partial x}. \tag{2.19}$$

Building on this formulation, a series of classical models such as the Payne model [77] and the Ross model [91] have been proposed, substantially advancing the field. Among them, the Payne model derives a dynamic equation for traffic flow—called a higher-order continuum model—by combining the continuity equation framework with intuitive reasoning from car-following models:

$$\frac{\mathrm{d}u}{\mathrm{d}t} = \frac{\partial u}{\partial t} + u\frac{\partial u}{\partial x} = \frac{u_e - u}{T} - \frac{u}{kT}\frac{\partial k}{\partial x}. \tag{2.20}$$

Here, the term $-\frac{u}{kT}\frac{\partial k}{\partial x}$ captures the expected response of following vehicles to upstream traffic conditions (i.e., density), while the relaxation term $\frac{u_e - u}{T}$ reflects the adjustment of speed towards an equilibrium speed u_e. This model allows for deviations from the equilibrium speed–density relationship, can describe the nonlinear wave propagation underlying congestion formation, and is able to analyze traffic instability under perturbations, stop-and-go waves, and phase transitions between traffic states. Compared to first-order models, higher-order models achieve stronger expressive power and higher accuracy in describing real traffic dynamics.

Mechanism-based spatio-temporal prediction methods offer clear physical interpretability and produce results that are explicitly consistent with physical laws. However, they also face several important limitations in practical applications. First, most existing mechanism-based models are constructed under idealized assumptions (e.g., closed systems, linear relationships) and are primarily suitable for simulating static or quasi-static scenarios. In contrast, real spatio-temporal data are typically highly dynamic and nonlinear, involving the interaction of multiple factors. This complexity makes it difficult for traditional mechanism-based models to accurately capture real-world spatio-temporal evolution, thereby limiting their predictive accuracy.

Second, the construction of mechanism-based models is heavily knowledge-dependent. For complex systems such as urban traffic flow or atmospheric pollution diffusion, our current scientific understanding is far from sufficient to comprehensively characterize all interaction mechanisms. This inherent knowledge gap fundamentally constrains the generality and scalability of mechanism-based models.

2.1.2 Deep Learning-Based Methods

With the rapid progress of deep learning, its ability to automatically learn complex patterns from large-scale data has attracted considerable interest in the spatio-temporal prediction community. Consequently, a large number of deep learning-based spatio-temporal prediction methods have been proposed (Table 2.1). The development of this line of work can be roughly divided into three stages: (i) an exploratory stage characterized by diverse model designs, (ii) a consolidation stage in which common modeling paradigms emerged, and (iii) a stage of innovative exploration that builds on these paradigms. The remainder of this section reviews representative methods in each stage.

2.1.2.1 Diversified Development Phase

The initial stage of deep learning-based spatio-temporal prediction attracted not only researchers in computer science, but also domain experts in transportation, meteorology, and other related fields. Because researchers from different backgrounds interpret spatio-temporal data in different ways, the early literature exhibits a highly diversified landscape, hence the term “period of diverse exploration”.

Some studies directly adopt multilayer neural networks for prediction. For example, [66] introduced a deep belief network (DBN) to model the spatio-temporal evolution of traffic speed and predict future speed changes. Subsequent work extended this line of research by refining the DBN architecture [101, 117], exploiting autoencoders [11], and employing fully connected networks [113].

Other studies reformulate spatio-temporal prediction as multiple univariate or multivariate time-series forecasting tasks. For instance, [68] was among the first to employ long short-term memory (LSTM) networks for traffic speed prediction, effectively addressing the long-range dependency problem and automatically determining the optimal temporal window. Building on this idea, some studies employ multiple LSTM modules for joint modeling to predict human mobility or traffic patterns over several future steps, often under a multi-task learning framework [100]. A large

Table 2.1 Summary of deep learning-based spatio-temporal forecasting methods

Diverse exploration	Direct prediction with NNs	DBN/Autoencoder/ Fully-connected networks	[66, 101, 117]/[11]/[113]
	Decompose into multiple time-series forecasts	Recurrent neural networks (RNN)	[47, 51, 68, 100, 125, 136, 145]
	Transform data into matrices	Standard CNN	[114, 139]
		3D CNN	[9, 58]
Paradigm consolidation	Temporal dependency modeling	Manually constructed multi-scale temporal dependencies	[34, 138, 141]
		Automatically learned temporal dependencies via deep networks	RNN-based [3, 17, 59, 144]
			TCN-based [34, 122, 123, 134]
	Spatial dependency modeling	Grid-based modeling of spatial locality	CNN-based [9, 35, 138]
		Graph-based modeling of non-Euclidean topologies	GNN-based [9, 31, 33, 59, 134]
Innovative exploration	Dynamic dependency modeling	Attention-based dynamic spatiotemporal dependencies	[123, 131]
	Spatiotemporal heterogeneity modeling	Static parameter/model allocation for different spatiotemporal types	[3, 75, 137]
		Meta-learning for dynamically generated parameters per spatiotemporal location	[26, 74, 130, 132]
		Self-supervised learning for dynamic representations	[41]

body of subsequent work follows this direction [47, 51, 125, 136, 145], introducing architectural variations while retaining LSTM as the core building block.

In addition, some studies apply convolutional neural networks (CNNs) to spatio-temporal modeling. A common strategy is to unfold the spatial dimension of spatio-temporal data into a two-dimensional "time-space" matrix, interpret it as an image, and then use standard CNNs for prediction [114, 139]. Benefiting from the computational efficiency of CNNs, many subsequent works adopt CNN-based architectures as the backbone for spatio-temporal prediction [9, 58].

Overall, this stage witnessed a proliferation of innovative models. Researchers systematically explored diverse deep learning paradigms which, despite their heterogeneity, played a crucial role in laying the methodological foundations and accumulating practical experience for deep learning-based spatio-temporal prediction (Table 2.1).

2.1.2.2 Paradigm Consolidation Phase

The second stage of research is characterized by a gradual consensus: spatiotemporal data should be modeled separately along the temporal and spatial dimensions. That is, temporal and spatial dependencies are supposed to be captured by dedicated architectural components.

Temporal Dependency Modeling

Temporal dependency modeling has formed two major branches: manual multi-scale modeling and automatic deep modeling.

For manual multi-scale modeling, researchers analyze spatiotemporal data and explicitly construct dependencies at multiple temporal scales [34, 138, 141]. As illustrated in Fig. 2.1, these dependencies typically include short-term closeness, medium-term periodicity, and long-term trend.

Closeness refers to the autocorrelation among nearby time steps. For example, a traffic accident at 9:10 AM may affect traffic conditions at 9:20 AM and even 9:30 AM. Figure 2.1a visualizes the ratio between samples at two time points as a function of their time lag (horizontal axis: time lag, vertical axis: average ratio). The ratio is close to 1 when the lag is within 2 hours, and increases as the time lag grows, indicating strong short-term temporal closeness.

Periodicity arises from the regularity of human daily activities, typically manifesting at daily or weekly scales; for instance, morning peak hours recur on workdays, as shown in Fig. 2.1b.

Trend captures long-term monotonic changes; for example, as the season changes from winter to summer, rising temperatures may cause the morning peak to occur earlier, leading to a decreasing trend in residential traffic volumes from 9:00 to 9:30 AM, as in Fig. 2.1c.

The second branch of work focuses on automatically learning temporal dependencies via deep neural networks, mainly recurrent neural networks (RNNs) and temporal convolutional networks (TCNs). RNN-based models are typically instantiated using long short-term memory (LSTM) units or gated recurrent units (GRUs). LSTMs

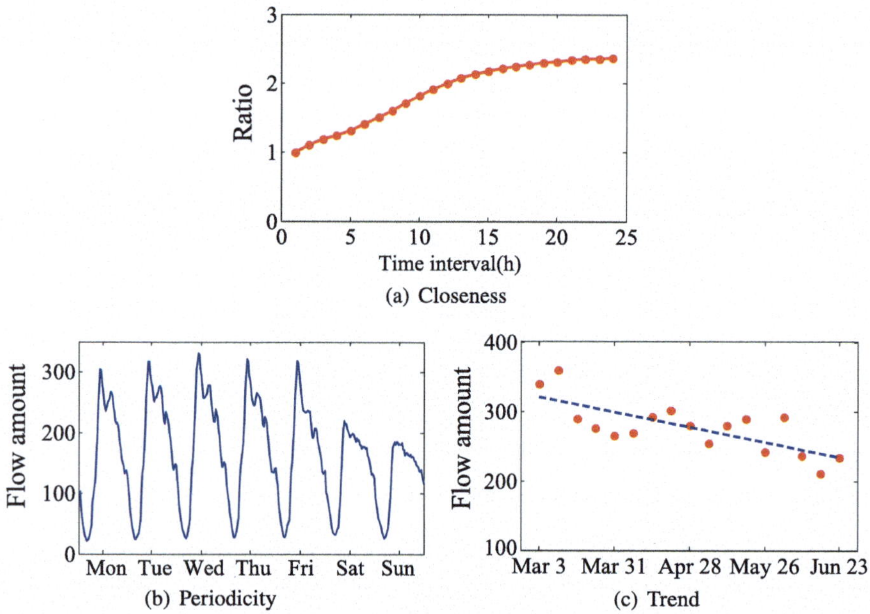

Fig. 2.1 Illustration of temporal dependencies at different scales

are effective in capturing long-range temporal dependencies, whereas GRUs offer a more compact parameterization with comparable performance. Representative spatio-temporal prediction models that use LSTM/GRU to model temporal dependencies include TGCN [144], TGCLSTM [17], DCRNN [59], and AGCRN [3]. Both DCRNN and AGCRN employ GRU as the backbone for modeling inter-step temporal dependencies, but differ architecturally: DCRNN adopts an encoder-decoder structure, whereas AGCRN uses an encoder-only design for prediction.

Unlike RNNs, TCNs are variants of one-dimensional CNNs that do not rely on the output from the previous time step, enabling parallel computation and thus improving efficiency in temporal modeling. Representative TCN-based spatio-temporal prediction models include STGCN [134] and ASTGCN [34]. Some works further introduce dilated causal convolutions [135], which allow the temporal receptive field to grow exponentially with network depth and thereby enhance the ability to capture long-range temporal dependencies. Representative models in this category include Graph WaveNet [123] and MTGNN [122].

Spatial Dependency Modeling

Spatial dependency modeling can be broadly divided into grid-based and graph-based approaches.

Early work typically uses standard 2D CNNs to capture spatial locality. A common approach is to partition the study area into a regular grid, construct a spatial matrix, and apply 2D CNNs to model spatial adjacency [138]. Some studies further stack the temporal dimension with the spatial grid to form a 3D spatiotemporal tensor, and

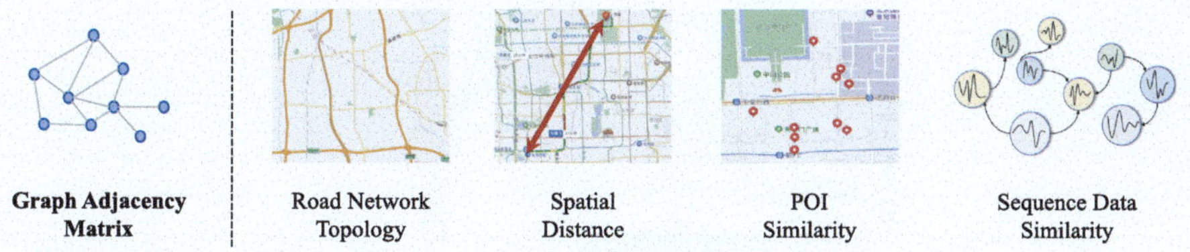

Fig. 2.2 Illustration of graph adjacency matrices and their construction

then use 3D CNNs to capture spatiotemporal locality, achieving favorable empirical performance [9, 35].

However, many spatiotemporal datasets naturally reside on non-Euclidean structures, such as urban traffic networks (road networks, metro networks) and meteorological monitoring station networks. These structures are better represented as graphs. Traditional CNNs struggle to process such non-Euclidean data, whereas graph neural networks (GNNs) are specifically designed for this purpose. Therefore, most recent work adopts GNNs for spatial dependency modeling. A key design choice in GNN-based models is how to construct the graph adjacency matrix. Existing methods include defining adjacency according to road network topology [59, 134], according to physical distance or similarity of point-of-interest (POI) functionality [9, 14, 31], and learning adaptive adjacency matrices directly from data [33, 122, 123], as illustrated in Fig. 2.2.

During this period, the research paradigm in spatiotemporal prediction gradually converged: temporal and spatial dependencies are modeled by appropriate deep architectures, respectively. This paradigm provided a solid foundation for subsequent research and significantly advanced the field.

2.1.2.3 Innovative Exploration Phase

After roughly five years of development under this unified paradigm, it became evident that the performance of spatio-temporal prediction models was approaching a plateau. Consequently, the research focus began to shift toward the intrinsic properties of the data themselves, seeking novel perspectives for innovation. This transition marks the third stage of research, in which two topics have attracted particular attention: dynamically evolving spatio-temporal dependencies [123, 131] and spatio-temporal heterogeneity [3, 75, 132, 137].

Dynamic spatio-temporal dependencies arise along both the temporal and spatial dimensions. Temporally, dependencies exhibit drifting periodic patterns. For example, morning peak hours may shift from day to day, so traffic flow does not strictly follow fixed periodic cycles. To address this issue, some studies introduce attention mechanisms that assign dynamic weights to historical time steps and perform weighted aggregation when generating predictions [131]. Spatially, region-to-region correlations also vary over time. For instance, correlations between residential areas

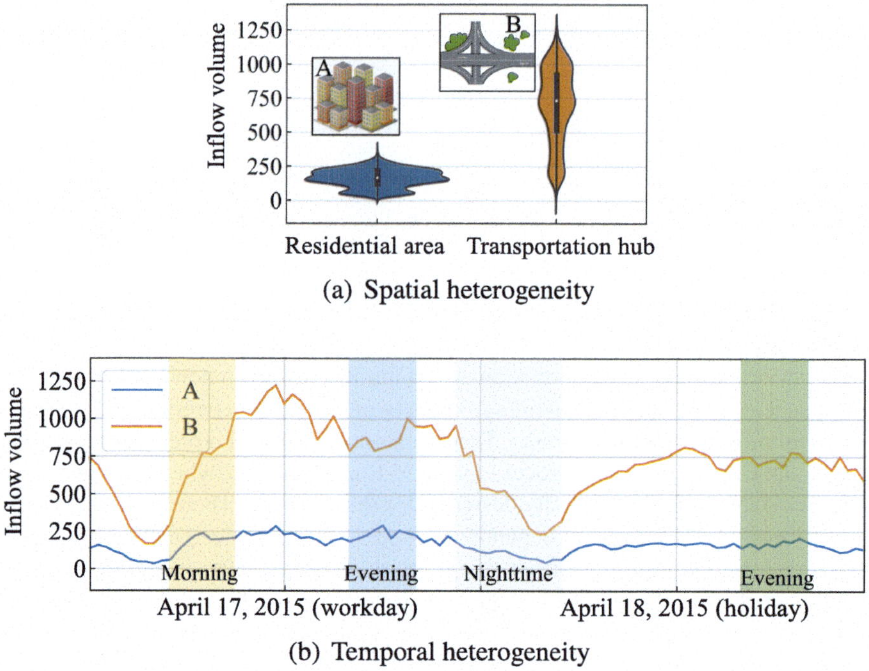

Fig. 2.3 Illustration of spatiotemporal heterogeneity

and business districts are strong during peak hours but much weaker during regular working hours. To capture such dynamic spatial dependencies, some work learns time-varying region similarities directly from data [123].

Spatiotemporal heterogeneity is also prominent along both spatial and temporal axes. As shown in Fig. 2.3, consider two functional regions in a city—a residential area (A) and a transportation hub (B). Figure 2.3a visualizes the distributions of their traffic flows, and Fig. 2.3b shows how traffic volumes evolve over time. We observe that spatiotemporal data exhibit heterogeneous patterns across space and time. Spatially, the residential area and transport hub have markedly different distributions in Fig. 2.3a; temporally, rush hours and off-peak periods display distinct patterns in Fig. 2.3b. This indicates that spatiotemporal data are characterized by both spatial and temporal heterogeneity. If a single model with shared parameters is used to fit data across all heterogeneous regions and time periods, it is difficult to attain optimal performance. This has motivated dedicated studies on spatial and temporal heterogeneity.

For spatial heterogeneity, the most straightforward approach is to assign distinct parameters or models to different regions, thereby capturing region-specific patterns. This idea has been validated in the literature [137]. However, it relies on the implicit assumption that the types of heterogeneous regions are known a priori, which limits its applicability when region types are unknown. Other work proposes a "one-model-

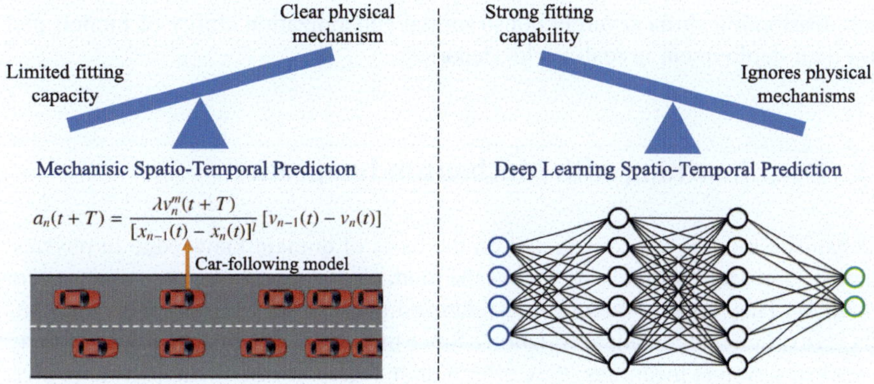

Fig. 2.4 Comparison between two modeling paradigms for spatiotemporal prediction

per-location" strategy, allocating an independent parameter space to each region [3]. This approach, however, leads to an explosion in the number of parameters. To alleviate this issue, matrix factorization techniques have been incorporated, yielding promising results [75].

While spatial heterogeneity is relatively fixed within a dataset, temporal heterogeneity is inherently dynamic, making it more challenging to model. To tackle this, recent work has adopted meta-learning [85], generating specialized initial parameters for different temporal or spatial patterns and thereby capturing spatiotemporal heterogeneity more effectively [26, 74, 130, 132]. Other studies attempt to model temporal heterogeneity by partitioning the parameter space [56, 99], but they typically assume that temporal heterogeneity is static over time, which often does not hold in practice. To address this, some researchers propose alternative strategies based on self-supervised learning: by maintaining dedicated dynamic representations for different time periods, temporal heterogeneity can be encoded directly into learned representations [41].

Thanks to sustained efforts in the community, deep learning-based models have achieved strong performance on standard benchmark datasets for spatiotemporal prediction. Nevertheless, they still suffer from two key limitations.

First, many spatiotemporal prediction applications require models not only to provide accurate forecasts but also to reveal the underlying generation and evolution mechanisms of spatiotemporal processes. However, deep models typically focus on minimizing prediction error and are not explicitly designed to encode domain mechanisms, as illustrated in Fig. 2.4. Consequently, they exhibit limited explanatory power regarding the underlying mechanisms of spatiotemporal dynamics.

Second, the prevailing design paradigm uses complex "black-box" models to approximate observed spatiotemporal samples (training data), with the expectation that they will generalize to unseen test data. In practice, the generation of spatiotemporal data is influenced by numerous external factors (e.g. weather conditions, large-scale events), which can induce distribution shifts between training and test sets.

Such distribution shifts severely constrain the generalization ability of models and limit their deployment in real-world scenarios.

2.2 Deep Learning with Mechanism Integration

Mechanistic models are constructed on the basis of domain knowledge in physics, chemistry, and related fields, as well as the fundamental principles governing system dynamics. They encode the intrinsic relationships between system inputs and outputs via explicit mathematical equations. Since the parameters in such models usually have clear physical meanings, they offer a high degree of interpretability. However, the construction of mechanistic models is constrained by the completeness of physical knowledge and by computational complexity, and typically requires accurate parameter estimation and carefully specified initial conditions. As a consequence, their development and validation often demand substantial time and resources.

In contrast, non-mechanistic models represented by deep learning optimize large numbers of weight parameters directly on large-scale datasets via gradient descent. These models can rapidly establish implicit mappings between inputs and outputs with relatively modest prior knowledge requirements. However, due to the lack of explicit encoding of internal mechanisms, such models are comparatively weak in interpretability, and their generalization ability is limited. Furthermore, the predictive performance of non-mechanistic models is highly dependent on data quality and the choice of fitting procedures. A schematic comparison between mechanistic modeling and deep learning paradigms is shown in Fig. 2.5.

Given the complementary strengths and limitations of mechanistic models and deep learning, there has been sustained interest in integrating these two paradigms. It has been repeatedly argued that future modeling frameworks should combine mechanistic modeling with machine learning techniques [86]. As an important branch of this research direction, deep learning methods that explicitly incorporate mechanistic principles have recently attracted extensive attention and application.

Representative examples include physics-informed neural networks and neural operators. These methods inject process-based mechanisms into data-driven models and exemplify the growing trend of combining mechanistic structure with machine learning. In doing so, they aim to retain the computational efficiency of deep learning while achieving higher levels of interpretability and generalization, thereby providing more robust solutions for scientific computing and engineering applications. This section reviews these two representative classes of methods, summarizing their advances and limitations.

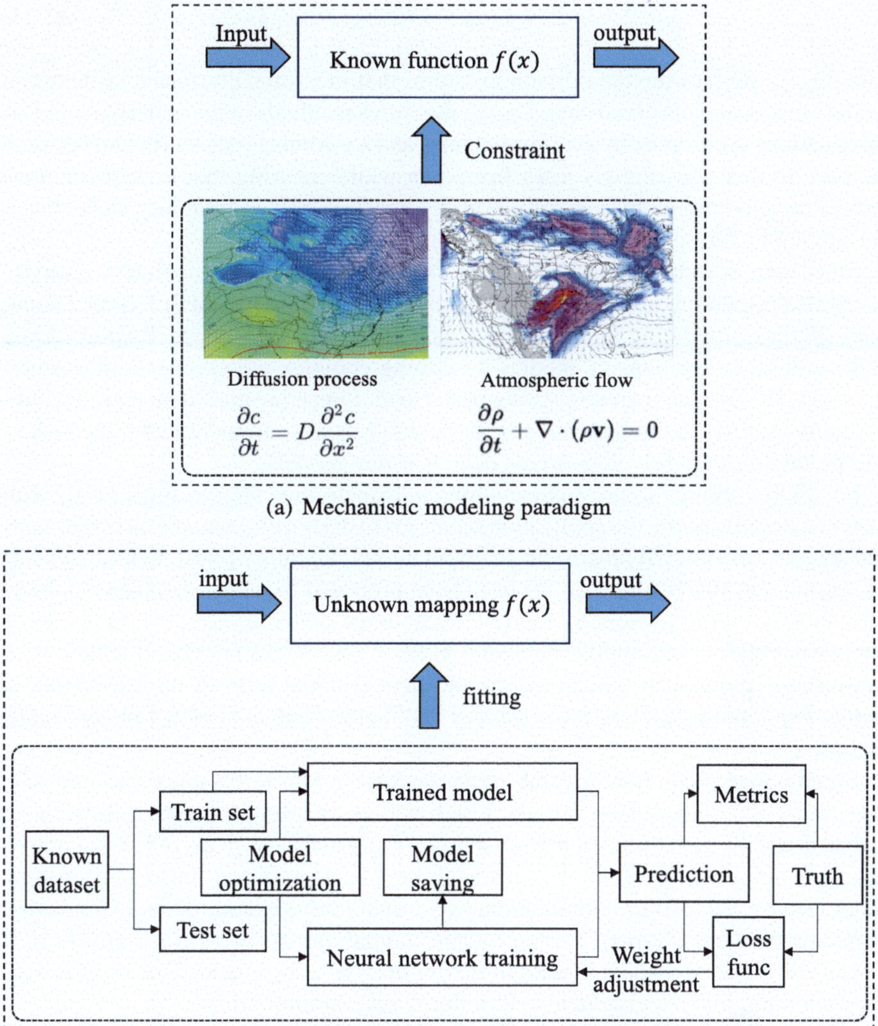

(a) Mechanistic modeling paradigm

(b) Deep learning modeling paradigm

Fig. 2.5 Modeling paradigms of mechanistic models and deep learning

2.2.1 *Physics-Informed Neural Networks*

Physics-informed neural networks (PINNs) have emerged as a prominent topic at the intersection of scientific computing and machine learning. Their core idea is to incorporate physical equations (e.g. conservation laws, Navier-Stokes equations) into the design of neural network loss functions or architectures, thereby achieving an organic integration of data-driven deep learning with physical constraints from mechanistic models. A general formulation can be written as

$$\mathcal{L} = \mathcal{L}_{\text{data}} + \lambda \mathcal{L}_{\text{physics}}, \tag{2.21}$$

where $\mathcal{L}_{\text{data}}$ denotes the data-driven loss term that measures discrepancies between model outputs and observed data, $\mathcal{L}_{\text{physics}}$ measures residuals of the underlying physical equations to enforce physical consistency, and λ is a balancing coefficient between the two. In this way, PINNs learn from data while ensuring that their predictions respect the governing physical laws, thereby improving interpretability and enhancing suitability for modeling complex physical systems.

Since their introduction in 2019 [84], PINNs have been widely applied to a variety of physical modeling and simulation tasks, such as solving partial differential equations (PDEs), simulating fluid dynamics, and predicting radio maps. By embedding PDEs and other mechanistic models describing complex environments into neural networks, PINNs can maintain strong performance even in small-data regimes, substantially improve computational efficiency, and partially compensate for the lack of interpretability in purely data-driven deep learning models.

For PDE solving, some studies combine Transformer architectures [110] with PDE constraints in specific application scenarios to design efficient mesh-free numerical solvers, substantially increasing computational flexibility [146]. In fluid dynamics, variational PINN formulations have been proposed to reduce reliance on observational data while improving numerical accuracy and stability [45]. Other work integrates PINNs with multiscale neural networks to obtain numerical solutions of atmospheric dynamical equations; the physical residual term in the PINN loss is further exploited to analyze error sources and thereby enhance model interpretability [64].

In radio map prediction, several studies transform Maxwell's equations into suitable forms and embed them into the loss function, enabling accurate estimation of path loss and the generation of high-resolution radio path-loss maps [42, 54]. For particle system simulation, PINN-based models have been proposed that exhibit stronger generalization and lower computational cost than traditional numerical simulators, providing efficient solutions for large-scale simulations of complex systems [93].

Despite the remarkable progress of PINNs in combining mechanistic models with deep learning, this integration also introduces several limitations.

- First, incorporating physical residual terms substantially increases the non-convexity of the optimization problem. The computation of high-order derivatives can lead to numerical instability during training; in particular, when modeling complex flows with shocks and discontinuities, the network can easily become trapped in poor local minima, hampering convergence and predictive accuracy.
- Second, for high-dimensional spatio-temporal problems, the heavy reliance of PINNs on high-order automatic differentiation causes the computational cost of the loss function to grow rapidly with dimensionality. This poses a severe computational challenge for large-scale industrial applications, making it difficult to directly scale PINNs to highly complex systems.
- Third, when dealing with multi-scale and multi-physics problems, PINNs still struggle in terms of training efficiency and generalization. A key reason is the dif-

ficulty of establishing an efficient and unified mapping between network parameters and the governing physical equations. The lack of such a mapping limits the applicability of PINNs to complex multi-physics coupling problems, especially those involving heterogeneous mechanisms and multi-scale features.

These limitations motivate the search for more efficient alternatives that can overcome the shortcomings of PINNs and further improve their feasibility and performance in real-world applications.

2.2.2 *Neural Operators*

Neural operators (NOs)[1] are fundamentally rooted in operator learning theory. Mathematically, the problem is formulated as follows: given paired observations of functions, where one function belongs to an input function space and the other to an output function space, the goal is to learn an approximation of an unknown operator that maps inputs to outputs. The learned operator should be able to predict output functions that closely match the ground truth for any new input function. Typical examples of such operators include differential operators, Laplacian operators, and other mappings arising in mathematical physics.

From the perspective of mechanistic modeling, the target operator often corresponds to the core mathematical mapping that characterizes the relationship between system inputs and outputs. However, in many complex application scenarios, the governing equations of the system are difficult to specify explicitly. By approximating these operators with neural networks, neural operators circumvent the need for fully explicit equations, thereby providing an effective alternative pathway for modeling complex systems.

The Fourier neural operator (FNO) [60] is one of the most representative neural-operator architectures. Its core idea is to replace the conventional message-passing mechanism in graph neural networks with Fourier and inverse Fourier transforms. Concretely, an FNO layer first applies a Fourier transform to the input, filters out high-frequency modes to suppress noise and improve robustness, and then applies an inverse Fourier transform to reconstruct the signal from the retained low-frequency modes. The resulting representation is combined with the original input through a residual connection to produce the final layer output. Subsequent work has proposed variant architectures such as U-FNO [118] and F-FNO [106], which embed U-Net modules [90] into FNO layers or apply dimension-wise Fourier transforms, respectively. These extensions further enhance the performance and applicability of FNO while preserving its core advantages.

Neural operators have been applied primarily to PDE solving, including the Navier-Stokes equations, advection equations, and other PDEs describing fluid flow and related physical processes. On benchmark platforms such as PDEBench [104],

[1] An introduction to neural operators can be found at https://en.wikipedia.org/wiki/Neural_operators.

the Fourier neural operator and its variants often outperform competing methods, demonstrating notable advantages in efficiently solving complex physical equations.

Compared with PINNs, neural operators are not tied to explicit physical formulas for specific scenarios and benefit from the flexibility of operator learning. They often exhibit stronger generalization ability and higher computational efficiency. However, the modeling paradigm of neural operators is more heavily data-driven and therefore lacks explicit physical constraints. As a result, their performance can degrade substantially when data are scarce, and in some settings they may even produce predictions that violate known physical laws.

2.3 Modeling Spatio-Temporal Distribution Shifts

Although existing spatio-temporal prediction methods have achieved impressive results, most studies rely on a basic assumption: training and test data are independently and identically distributed, i.e., both are assumed to be independently sampled from the same underlying distribution. In practice, however, spatio-temporal prediction often faces distribution shift due to exogenous factors such as weather, special events, or policy interventions, so the distribution of spatio-temporal data can differ substantially between training and testing phases.

Spatio-temporal data inherently possess both temporal and spatial attributes. Accordingly, this section reviews distribution-shift modeling from three perspectives: the temporal dimension, the spatial dimension, and the joint spatio-temporal dimension. Because research on distribution shift in spatio-temporal prediction is still relatively limited, our discussion of temporal and spatial distribution shift focuses mainly on methods developed for time-series data and graph data, respectively.

2.3.1 Modeling Along Temporal Dimensions

Temporal distribution-shift modeling aims to address changes in the data-generating process of time-series signals over time. Time series are widely used in domains such as stock prediction, weather forecasting, user activity prediction, electricity demand forecasting, and medical analytics [52, 69, 147, 148]. In real applications, the statistical properties of time series often evolve, leading to pronounced non-stationarity that manifests as dynamic distribution shift. For example, in stock prediction, variations in market volatility across different periods can induce substantial shifts between the distributions of past and future financial factors [49].

Early work proposed hidden Markov models [29], RNN-based models [40], and Transformer-based models [57] for time-series forecasting, achieving notable performance gains. However, distribution shifts between training and test sets, together with unobserved latent factors that influence data distributions, generally violate the i.i.d. assumptions underlying these models and thus limit their generalization ability.

To tackle temporal distribution shift, existing research has explored three main directions: causal learning, feature decoupling, and distribution matching.

In the *causal learning* direction, some studies argue that temporal distribution shift primarily arises because external factors affecting event occurrence are not explicitly modeled, and refer to these external factors as “contexts” [128]. They analyze temporal distribution shift from the perspective of data generation. Specifically, structural causal models (SCMs) are used to capture causal dependencies among contexts, historical sequences, and labels, thereby characterizing the structural properties of data generation and prediction. In addition, variational context-adjustment methods are proposed to uncover latent associations between contexts and observed sequences, guiding models to focus on critical causal pathways and improving robustness and adaptability to temporal distribution shift.

In the *feature decoupling* direction, some work draws on ideas from structured time-series models [81], arguing that learning decoupled seasonal and trend representations is more advantageous for forecasting [120]. Feature decoupling can better accommodate distribution shift and thereby improve robustness and generalization. It has been pointed out that end-to-end neural models are prone to capture spurious correlations and learn entangled representations; when distribution shift occurs in a specific component, such models are difficult to adapt. To alleviate this issue, a mixture-of-autoregressive-experts architecture and learnable Fourier layers are introduced to separately model seasonal and trend components, and contrastive learning is employed to decouple features, thereby effectively addressing distribution shift in time-series forecasting.

In the *distribution matching* direction, some studies formally define the temporal covariate-shift problem and propose an adaptive RNN (AdaRNN) to address it [24]. From the standpoint of distributional representation, AdaRNN designs a temporal distribution characterization module to encode distributional information in time series, together with a temporal distribution-matching module to mitigate distribution mismatch. With these two modules, the adaptive RNN can dynamically adjust model parameters to accommodate distribution changes over time, thus improving robustness to temporal covariate shift. Another line of work proposes a general framework called DIVERSIFY [65] for out-of-distribution detection and generalization in time-series data. It emphasizes that, in computer vision, distribution-shift modeling often assumes multiple domains with static and known labels [28, 116], whereas in time-series data, distributions evolve dynamically and may comprise multiple latent subdomains. To handle this, DIVERSIFY leverages adversarial training to segment time series into multiple latent domains and learns domain-invariant representations by reducing distribution discrepancies among them, thereby effectively addressing distribution shift in time-series forecasting.

2.3.2 Modeling Along Spatial Dimensions

In recent years, neural networks for graph-structured data—such as graph neural networks (GNNs) [94], graph convolutional networks (GCNs) [20], and graph attention networks (GATs) [111]—have achieved remarkable success on tasks involving social networks, molecular graphs, and knowledge graphs.

However, most existing studies do not explicitly account for distribution shift in graph-structured data. Such shifts mainly manifest as node feature shift and graph structure shift. Node feature shift refers to changes in the distribution of node attributes across domains, whereas graph structure shift is reflected in differences in graph size, subgraph density, and link patterns across datasets [6, 133].

Due to the non-Euclidean nature of graph data, the interdependence between edges and node features, and the susceptibility of GNNs to spurious correlations [108], modeling spatial distribution shift with GNNs is challenging.

Existing approaches primarily solve this problem from three angles: graph data augmentation, causal inference, and representation disentanglement.

In *graph data augmentation*, methods can be broadly classified into rule-based and model-based approaches. Among rule-based methods, DropEdge [89] randomly removes a fraction of edges at each training epoch to introduce structural noise and enhance robustness, while AdaEdge [10] iteratively adds edges between nodes with highly similar labels to enrich structural information. Although these methods augment data effectively, they are susceptible to error propagation, which can lead to unstable performance.

Among model-based methods, BGCN [142] iteratively trains a mixture-membership stochastic block model guided by GCN predictions to generate multiple denoised graphs, and ensembles the corresponding GCN outputs to improve generalization. GAUG [67] employs a neural edge predictor to estimate edge existence probabilities and learn class-homophily trends; GAUG-M and GAUG-O then modify the original graph structure according to these predictions to generate augmented edges. MHAug [76] introduces an explicit target distribution and uses Metropolis-Hastings sampling to generate augmented graphs that follow this target, providing a probabilistic augmentation framework.

In the *causal inference* direction, CIGA [13] captures graph invariances from a causal perspective to guarantee generalization under distribution shift. Concretely, it constructs three structural causal models to characterize potential shifts on graphs and extends the invariance principle to the graph domain. An information-theoretic objective is then designed to extract subgraphs that preserve maximal label-invariant information. DIR [121] also introduces causal intervention and invariant learning. It comprises four modules: a rationale generator that splits the input graph into causal and non-causal subgraphs, a distributional interventional module that infers invariances in the causal components, a feature encoder, and two classifiers that jointly predict from causal and non-causal parts, thereby improving both generalization and interpretability.

Representation disentanglement methods aim to learn node and graph representations that separate underlying factors of variation, thereby enhancing robustness and out-of-distribution generalization [1]. DisenGCN [67] proposes a neighbor-routing mechanism that dynamically identifies latent factors underlying edges between neighbors and assigns neighbors to channels that extract and convolve factor-specific features, yielding an end-to-end disentangled graph convolutional network. OOD-GNN [55] introduces a nonlinear graph representation decorrelation method based on random Fourier features [61, 83, 140] and sample re-weighting, which helps remove spurious correlations and substantially improves out-of-distribution generalization. It also designs a scalable weight estimator to control computational cost and enhance efficiency.

2.3.3 Modeling Along Joint Spatio-Temporal Dimensions

The preceding subsections have reviewed distribution-shift modeling in the temporal and spatial dimensions separately, without jointly accounting for shifts in both. In contrast, spatio-temporal distribution-shift modeling focuses on shifts in spatio-temporal graph data.

Because spatio-temporal graphs exhibit intricate structure in both time and space, the modeling task must simultaneously track distributional changes along both dimensions and capture their dynamic interplay, i.e., the interaction between temporal and spatial dependencies. This renders the problem substantially more complex and challenging.

Existing work approaches spatio-temporal distribution shift mainly from three directions: causal inference, invariance learning, and neural processes.

In terms of *causal inference*, CaST [124] constructs a structural causal model for spatio-temporal graphs to analyze the data-generation process from a causal perspective. Based on this model, CaST applies back-door adjustment to handle distribution shift along the temporal dimension and designs a decoupling module to separate input data into time-dependent and time-independent components. Furthermore, CaST uses front-door adjustment combined with edge-level convolutional operators to capture dynamic causal relationships between nodes, thereby achieving spatio-temporal distribution-shift modeling. STNSCM [23] also builds a causal model, but employs the front-door criterion to remove confounding bias during feature extraction. It further designs a counterfactual representation inference module that extrapolates spatio-temporal states from factual to counterfactual scenarios, and applies the resulting representations to bike-sharing demand prediction. CauSTG [150] studies spatio-temporal generalization from an invariance perspective by transforming spatio-temporal relationships into learnable parameters. It introduces a spatio-temporal consistency learner and a hierarchical invariance explorer to capture invariant features that mitigate the impact of heterogeneous risks on relational extraction. The consistency learner decouples spatial relationships into positively and negatively causal components and quantifies them with trainable parameters.

In terms of *invariance learning*, STONE [112] is the first work to systematically investigate out-of-distribution learning for spatio-temporal data while jointly considering temporal dynamics and structural changes. It employs gated Transformers [110] to construct spatial and temporal semantic graphs and thereby capture spatio-temporal heterogeneity. STONE further introduces Fréchet embeddings to encode graph topology and extract stable dependencies. In addition, a spatio-temporal graph intervention mechanism is proposed to perturb the two types of semantic graphs and generate multiple perturbed environments, enabling the model to mine invariant node representations without relying on explicit data augmentation.

In terms of *neural-process* methods, STGNP [38] proposes a neural latent-variable model that combines neural networks with neural processes to capture complex spatio-temporal correlations while providing uncertainty estimates. Specifically, STGNP stacks causal convolution layers and cross-set GNN layers to learn deterministic spatio-temporal representations, and then models latent variables at target locations via vertical latent state transitions between layers, thereby enabling extrapolative prediction. It further designs a Bayesian graph aggregator that aggregates contextual information while accounting for uncertainties in both context data and graph structures.

2.4 Spatio-Temporal Decomposition Prediction

Spatiotemporal decomposition-based prediction methods are built on the idea of decomposing raw spatiotemporal data into multiple components with simpler spatiotemporal patterns and then modeling and forecasting each component separately. By reducing data heterogeneity and enhancing pattern separability, this paradigm provides an effective route for spatiotemporal prediction.

This section first introduces the underlying ideas and origin of decomposition-based methods, and then reviews two typical paradigms: two-stage and end-to-end spatiotemporal decomposition prediction methods.

2.4.1 *Origin of Decomposition Prediction*

The theoretical roots of decomposition-based prediction can be traced back to classical time-series decomposition methods [78]. This paradigm builds on the understanding that time series often exhibit multiple underlying patterns and is regarded as an effective approach for time-series analysis.

The core idea is to use decomposition techniques to separate complex temporal patterns into subcomponents with clear physical meanings [115]. Formally, the goal of time-series decomposition is to decompose a non-stationary time series $\{X_t\}_{t=1}^{H}$ of length H into deterministic and stochastic components $\{Z_t\}_{t=1}^{H}$.

- The *deterministic* component is predictable and is estimated or extrapolated for forecasting; it typically includes a trend component $\{T_t\}_{t=1}^{H}$ that captures long-term changes and a seasonal component $\{S_t\}_{t=1}^{H}$ that captures periodic fluctuations.
- The *stochastic* component contains the remaining information beyond trend and seasonality, and analyzing it is crucial for improving prediction accuracy [18].

Depending on the type of time series, decomposition schemes vary in complexity, but they share a common goal of extracting the residual stochastic component $\{Z_t\}_{t=1}^{H}$.

Classical time-series decomposition methods predominantly adopt two paradigms: additive and multiplicative decomposition [62].

In the *additive* framework, the series is decomposed as

$$X_t = T_t + S_t + Z_t. \tag{2.22}$$

Representative methods include the X11 decomposition procedure [18] developed jointly by the U.S. Census Bureau and Statistics Canada, which uses moving averages to separate trend and seasonality; seasonal extraction methods based on ARIMA models, which effectively handle complex seasonal patterns [48]; and the STL method based on locally weighted regression, which is widely used for non-stationary time series due to its robustness and adaptivity [16].

Multiplicative decomposition, by contrast, assumes

$$X_t = T_t \times S_t \times Z_t. \tag{2.23}$$

For example, parametrized power transforms can stabilize variance by transforming non-constant variance series into constant-variance ones, thereby improving the linear modeling capacity and making the data distribution closer to normality [7]. These two paradigms preserve their respective strengths and have inspired hybrid decomposition models in which interactions between trend and seasonal components are captured via adaptive weighting mechanisms [37].

Classical time-series decomposition methods provide important inspiration for the design of spatiotemporal decomposition-based prediction. However, their extension to spatiotemporal data requires a deep understanding of spatiotemporal characteristics. Compared with univariate time series, spatiotemporal data differ fundamentally along three dimensions [2]: (i) spatial dependence, which induces correlations in data generation across neighboring regions (e.g. diffusion of temperature fields in meteorological systems); (ii) spatial heterogeneity, whereby different locations at the same time can follow distinct evolution patterns (e.g. different day–night traffic patterns in business vs. residential areas); and (iii) spatiotemporal coupling, whereby temporal dynamics can be modulated by spatial structures (e.g. how contact network topology shapes epidemic spread).

These properties do not invalidate the core idea of time-series decomposition; instead, they highlight the necessity of extending decomposition paradigms to higher-dimensional spaces. Depending on whether the optimization objectives of the decom-

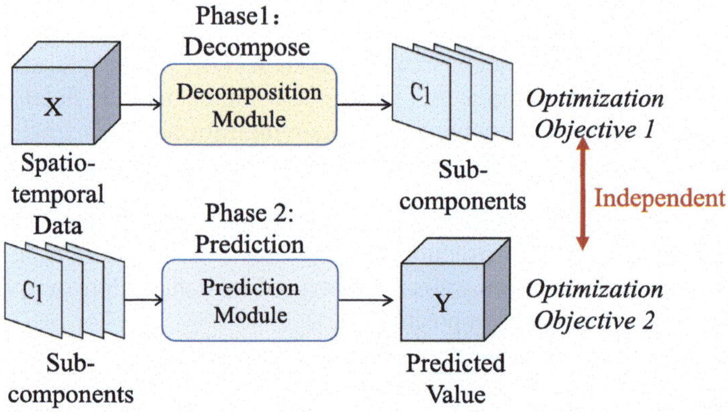

(a) Two-stage paradigm

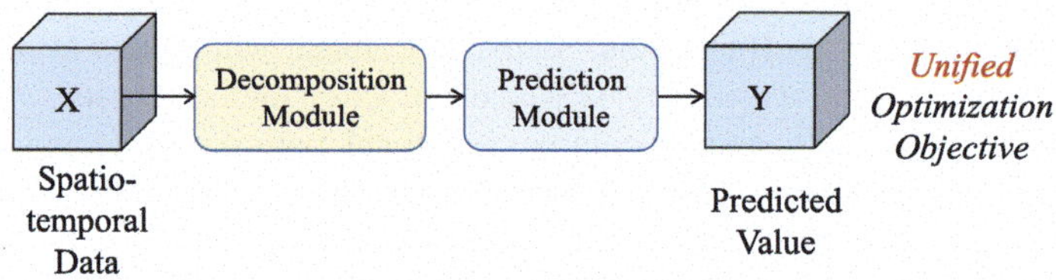

(b) End-to-end paradigm

Fig. 2.6 Two-stage and end-to-end spatiotemporal decomposition-based prediction paradigms

position and prediction stages are aligned, spatiotemporal decomposition-based methods can be divided into two-stage and end-to-end paradigms, as illustrated in Fig. 2.6.

2.4.2 Two-Stage Decomposition Prediction Methods

Two-stage spatio-temporal decomposition-based prediction methods adopt a hierarchical modeling framework consisting of a decomposition stage followed by a prediction stage. In the first stage, raw spatio-temporal data are preprocessed and decoupled into multiple subcomponents with explicit physical meaning. In the second stage, each component is modeled and forecast separately, and the two stages are optimized independently. The main theoretical advantage of this approach is that decomposition reduces modeling complexity while preserving the interpretability of individual components.

At the decomposition level, existing work mainly follows two technical routes. The first is rooted in multiscale decomposition theory from signal processing, which uses time-frequency analysis tools to extract spatio-temporal features at multiple scales. Representative applications include hybrid models that combine empirical mode decomposition (EMD) [82] with grey support vector machines, where intrinsic mode function decomposition is used for short-term traffic flow forecasting [43], and methods that adopt wavelet packet decomposition,[2] exploiting its multiresolution analysis capability to adaptively separate traffic-flow features [15]. Although these methods offer strong adaptivity, they require aggregating current and historical data for joint operations, which leads to substantial computational costs on large-scale datasets and creates efficiency bottlenecks in practice.

The second route is based on spectral analysis theory, which decomposes spatio-temporal data using predefined combinations of basis functions. Some work employs Fourier series expansion, using combinations of sine and cosine functions to model periodic components and constructing hybrid ARIMA-GARCH models to separately handle deterministic components and volatility residuals [143]. Other studies propose spatio-temporal decomposition methods based on locally weighted regression, using nonparametric regression to decompose traffic flow into trend and periodic components [149].

While such methods yield decompositions with clear physical interpretations, they rely on manually specified combinations of basis functions and lack self-adaptive capability for complex spatio-temporal patterns.

2.4.3 End-to-End Decomposition Prediction Methods

End-to-end spatio-temporal decomposition-based prediction methods adopt a joint optimization framework in which the decomposition process is tightly integrated with the prediction model. Unlike the two-stage paradigm, decomposition operators are implemented as differentiable modules embedded in neural networks and trained via backpropagation under an end-to-end objective. Depending on the underlying decomposition mechanisms, existing approaches can be grouped into two theoretical paradigms.

The first paradigm is diffusion-based, which decomposes spatio-temporal data into a diffusion component and an intrinsic component, feeds them into separate modules for diffusion and intrinsic-signal learning, and then combines the outputs via a gated fusion module [96]. This physically inspired signal-separation mechanism effectively models diffusion processes in spatio-temporal evolution.

The second line of work, grounded in spectral theory, adopts multi-scale decomposition strategies to process spatio-temporal signals. For instance, some studies decompose cellular traffic data into three sub-components—trend, periodicity, and randomness—and then perform causal structure learning based on variable-lag transfer entropy, modeling each sub-component as a directed causal graph [12]. Other work, inspired by the N-BEATS framework, decomposes air-quality time series into four interpretable parts, namely bias, slope, residual, and seasonality [46]. In addition,

[2] An introduction to wavelet packet decomposition can be found at https://en.wikipedia.org/wiki/Wavelet_packet_decomposition.

STWave constructs a multi-level discrete wavelet transform architecture: it leverages multiresolution analysis to extract smooth trend terms and employs a high-frequency feature fusion mechanism to capture bursty event patterns. While preserving the dual-channel modeling advantage, the multi-stage decomposition enables a cleaner trend estimation and more accurate characterization of low- and high-frequency spatio-temporal dependencies under long- and short-term regimes [27]. Moreover, a four-component disentanglement framework is proposed to decompose spatio-temporal data into long-term trend, periodic fluctuations, short-term perturbations, and spatial co-evolution components; it suppresses high-frequency noise via frequency-domain filtering and designs dedicated modules to model the evolution dynamics of each component[22]. Overall, these methods retain the benefits of multi-channel modeling while using spectral analysis to explicitly separate signal constituents.

End-to-end spatio-temporal decomposition forecasting methods further integrate the decomposition procedure with the forecasting model, enabling more efficient parameter optimization and more accurate predictions. Existing studies are mainly developed along two theoretical lines, i.e. diffusion theory and spectral theory. The former introduces physics-inspired signal separation mechanisms, whereas the latter employs multi-scale decomposition to effectively model the complex dynamics of spatio-temporal data.

Beyond improving forecasting accuracy, these methods also offer new theoretical perspectives and technical pathways for spatio-temporal data analysis. Nevertheless, many of these approaches remain largely heuristic and lack rigorous theoretical underpinnings. In particular, how to design decomposition algorithms that can reliably improve the performance of the same forecasting backbone is still an open problem.

References

1. Alemi, A.A., Fischer, I., Dillon, J.V., Murphy, K.: Deep variational information bottleneck. In: Proceedings of the 5th International Conference on Learning Representations (ICLR'17), pp. 1–19 (2017)
2. Atluri, G., Karpatne, A., Kumar, V.: Spatio-temporal data mining: A survey of problems and methods. ACM Computing Surveys **51**(4), 1–41 (2018)
3. Bai, L., Yao, L., Li, C., Wang, X., Wang, C.: Adaptive graph convolutional recurrent network for traffic forecasting. In: Proceedings of the 34th International Conference on Neural Information Processing Systems (NeurIPS'20), pp. 17804–17815 (2020)
4. Barthélemy, M.: Spatial networks. Physics Reports **499**(1), 1–101 (2011)
5. Benekohal, R.F., Treiterer, J.: CARSIM: car-following model for simulation of traffic in normal and stop-and-go conditions. Transportation Research Record **1194**(1), 99–111 (1988)
6. Bevilacqua, B., Zhou, Y., Ribeiro, B.: Size-invariant graph representations for graph classification extrapolations. In: Proceedings of the 38th International Conference on Machine Learning (ICML'21), pp. 837–851 (2021)
7. Box, G.E., Cox, D.R.: An analysis of transformations. Journal of the Royal Statistical Society Series B: Statistical Methodology **26**(2), 211–243 (1964)
8. Brackstone, M., McDonald, M.: Car-following: A historical review. Transportation Research Part F: Traffic Psychology and Behaviour **2**(4), 181–196 (1999)

9. Chen, C., Li, K., Teo, S.G., Chen, G., Zou, X., Yang, X., Vijay, R.C., Feng, J., Zeng, Z.: Exploiting spatio-temporal correlations with multiple 3D convolutional neural networks for citywide vehicle flow prediction. In: Proceedings of the 2018 IEEE International Conference on Data Mining (ICDM'18), pp. 893–898 (2018)
10. Chen, D., Lin, Y., Li, W., Li, P., Zhou, J., Sun, X.: Measuring and relieving the over-smoothing problem for graph neural networks from the topological view. In: Proceedings of the 34th AAAI Conference on Artificial Intelligence (AAAI'20), pp. 3438–3445 (2020)
11. Chen, Q., Song, X., Yamada, H., Shibasaki, R.: Learning deep representation from big and heterogeneous data for traffic accident inference. In: Proceedings of the 30th AAAI Conference on Artificial Intelligence (AAAI'16), pp. 338–344 (2016)
12. Chen, X., Chuai, G., Zhang, K., Gao, W.: Spatial-temporal cellular traffic prediction: A novel method based on causality and graph attention network. In: Proceedings of the 2023 IEEE Wireless Communications and Networking Conference, pp. 1–6 (2023)
13. Chen, Y., Zhang, Y., Bian, Y., et al.: Learning causally invariant representations for out-of-distribution generalization on graphs. In: Proceedings of the 36th International Conference on Neural Information Processing Systems (NeurIPS'22), pp. 22131–22148 (2022)
14. Cheng, J., Wang, J., Zhang, Y., Ji, J., Zhu, Y., Zhang, Z., Zhao, X.: Poi-enhancer: An llm-based semantic enhancement framework for poi representation learning. In: Proceedings of the 39th AAAI Conference on Artificial Intelligence (AAAI'25), pp. 1–9 (2025)
15. Cheng, Z., Li, X., Bai, Y., Li, C.: Multi-scale fuzzy inference system for influent characteristic prediction of wastewater treatment. CLEAN–Soil, Air, Water **46**(7), 1700343 (2018)
16. Cleveland, R.B., Cleveland, W.S., Terpenning, I.: STL: A seasonal-trend decomposition procedure based on LOESS. Journal of Official Statistics **6**(1), 3 (1990)
17. Cui, Z., Henrickson, K., Ke, R., Wang, Y.: Traffic graph convolutional recurrent neural network: A deep learning framework for network-scale traffic learning and forecasting. IEEE Transactions on Intelligent Transportation Systems (ITS) **21**(11), 4883–4894 (2019)
18. Dagum, E.B., Bianconcini, S.: Seasonal Adjustment Methods and Real Time Trend-Cycle Estimation. Springer (2016)
19. Davis, K.F., D'Odorico, P., Laio, F., Ridolfi, L.: Global spatio-temporal patterns in human migration: A complex network perspective. PLOS One **8**(1), e53723 (2013)
20. Defferrard, M., Bresson, X., Vandergheynst, P.: Convolutional neural networks on graphs with fast localized spectral filtering. In: Proceedings of the 30th International Conference on Neural Information Processing Systems (NeurIPS'16), pp. 3844–3852 (2016)
21. Dejon, B.: Spatial interaction network flow models. In: Proceedings of the Annual Meeting 1977 DGOR, pp. 377–386 (1978)
22. Deng, J., Chen, X., Jiang, R., Yin, D., Yang, Y., Song, X., Tsang, I.W.: Disentangling structured components: Towards adaptive, interpretable and scalable time series forecasting. IEEE Transactions on Knowledge and Data Engineering (TKDE) **36**(8), 3783–3800 (2024)
23. Deng, P., Zhao, Y., Liu, J., Jia, X., Wang, M.: Spatio-temporal neural structural causal models for bike flow prediction. In: Proceedings of the 37th AAAI Conference on Artificial Intelligence (AAAI'23), pp. 4242–4249 (2023)
24. Du, Y., Wang, J., Feng, W., Pan, S., Qin, T., Xu, R., Wang, C.: AdaRNN: Adaptive learning and forecasting of time series. In: Proceedings of the 30th ACM International Conference on Information and Knowledge Management (CIKM'21), pp. 402–411 (2021)
25. Elefteriadou, L., et al.: An introduction to traffic flow theory. Springer (2014)
26. Fang, S., Pan, X., Xiang, S., Pan, C.: Meta-MSNet: Meta-learning based multi-source data fusion for traffic flow prediction. IEEE Signal Processing Letters **28**(1), 6–10 (2020)
27. Fang, Y., Qin, Y., Luo, H., Zhao, F., Xu, B., Zeng, L., Wang, C.: When spatio-temporal meet wavelets: Disentangled traffic forecasting via efficient spectral graph attention networks. In: Proceedings of the 2023 IEEE 39th International Conference on Data Engineering (ICDE'23), pp. 517–529 (2023)
28. Fang, Z., Li, Y., Lu, J., Dong, J., Han, B., Liu, F.: Is out-of-distribution detection learnable? In: Proceedings of the 36th International Conference on Neural Information Processing Systems (NeurIPS'22), pp. 37199–37213 (2022)

29. Fulcher, B.D., Jones, N.S.: Highly comparative feature-based time-series classification. IEEE Transactions on Knowledge and Data Engineering (TKDE) **26**(12), 3026–3037 (2014)
30. Gazis, D.C., Herman, R., Rothery, R.W.: Nonlinear follow-the-leader models of traffic flow. Operations Research **9**(4), 545–567 (1961)
31. Geng, X., Li, Y., Wang, L., Zhang, L., Yang, Q., Ye, J., Liu, Y.: Spatiotemporal multi-graph convolution network for ride-hailing demand forecasting. In: Proceedings of the 33rd AAAI Conference on Artificial Intelligence (AAAI'19), pp. 3656–3663 (2019)
32. Gipps, P.G.: A behavioural car-following model for computer simulation. Transportation Research Part B: Methodological **15**(2), 105–111 (1981)
33. Guo, K., Hu, Y., Sun, Y., Qian, S., Gao, J., Yin, B.: Hierarchical graph convolution network for traffic forecasting. In: Proceedings of the 35th AAAI Conference on Artificial Intelligence (AAAI'21), pp. 151–159 (2021)
34. Guo, S., Lin, Y., Feng, N., Song, C., Wan, H.: Attention based spatial-temporal graph convolutional networks for traffic flow forecasting. In: Proceedings of the 33rd AAAI Conference on Artificial Intelligence (AAAI'19), pp. 922–929 (2019)
35. Guo, S., Lin, Y., Li, S., Chen, Z., Wan, H.: Deep spatial-temporal 3D convolutional neural networks for traffic data forecasting. IEEE Transactions on Intelligent Transportation Systems (ITS) **20**(10), 3913–3926 (2019)
36. Helbing, D.: Traffic and related self-driven many-particle systems. Reviews of modern physics **73**(4), 1067 (2001)
37. Holt, C.C.: Forecasting seasonals and trends by exponentially weighted moving averages. International Journal of Forecasting **20**(1), 5–10 (2004)
38. Hu, J., Liang, Y., Fan, Z., Chen, H., Zheng, Y., Zimmermann, R.: Graph neural processes for spatio-temporal extrapolation. In: Proceedings of the 29th ACM SIGKDD Conference on Knowledge Discovery and Data Mining (KDD'23), pp. 752–763 (2023)
39. Hua, C.I., Porell, F.: A critical review of the development of the gravity model. International Regional Science Review **4**(2), 97–126 (1979)
40. Hüsken, M., Stagge, P.: Recurrent neural networks for time series classification. Neurocomputing **50**(1), 223–235 (2003)
41. Ji, J., Wang, J., Huang, C., Wu, J., Xu, B., Wu, Z., Junbo, Z., Zheng, Y.: Spatio-temporal self-supervised learning for traffic flow prediction. In: Proceedings of the 37th AAAI Conference on Artificial Intelligence (AAAI'23), pp. 4356–4364 (2023)
42. Jiang, F., Li, T., Lv, X., Rui, H., Jin, D.: Physics-informed neural networks for path loss estimation by solving electromagnetic integral equations. IEEE Transactions on Wireless Communications (TWC) **23**(1), 15380–15393 (2024)
43. Jiang, X., Zhang, L., Chen, X.M.: Short-term forecasting of high-speed rail demand: A hybrid approach combining ensemble empirical mode decomposition and gray support vector machine with real-world applications in China. Transportation Research Part C: Emerging Technologies **44**(1), 110–127 (2014)
44. Jin, G., Liang, Y., Fang, Y., Shao, Z., Huang, J., Zhang, J., Zheng, Y.: Spatio-temporal graph neural networks for predictive learning in urban computing: A survey. IEEE Transactions on Knowledge and Data Engineering (TKDE) **36**(10), 5388–5408 (2024)
45. Kharazmi, E., Zhang, Z., Karniadakis, G.E.: *hp*-VPINNs: Variational physics-informed neural networks with domain decomposition. Computer Methods in Applied Mechanics and Engineering **374**(1), 113547 (2021)
46. Kim, T., Kim, J., Yang, W., Lee, H., Choo, J.: Missing value imputation of time-series air-quality data via deep neural networks. International Journal of Environmental Research and Public Health **18**(22), 12213 (2021)
47. Kong, D., Wu, F.: HST-LSTM: A hierarchical spatial-temporal long-short term memory network for location prediction. In: Proceedings of the 27th International Joint Conference on Artificial Intelligence (IJCAI'18), pp. 2341–2347 (2018)
48. Kumar, S.V., Vanajakshi, L.: Short-term traffic flow prediction using seasonal ARIMA model with limited input data. European Transport Research Review **7**(3), 1–9 (2015)

49. Kuznetsov, V., Mohri, M.: Generalization bounds for time series prediction with non-stationary processes. In: Proceedings of the International Conference on Algorithmic Learning Theory, pp. 260–274 (2014)
50. Kwan, M.P.: Mobile communications, social networks, and urban travel: Hypertext as a new metaphor for conceptualizing spatial interaction. The Professional Geographer **59**(4), 434–446 (2007)
51. Laptev, N., Yosinski, J., Li, L.E., Smyl, S.: Time-series extreme event forecasting with neural networks at Uber. In: Proceedings of the 34th International Conference on Machine Learning (ICML'17), pp. 1–5 (2017)
52. Le Guen, V., Thome, N.: Shape and time distortion loss for training deep time series forecasting models. In: Proceedings of the 33rd International Conference on Neural Information Processing Systems (NeurIPS'19), pp. 4189–4201 (2019)
53. Lee, G.: A generalization of linear car-following theory. Operations Research **14**(4), 595–606 (1966)
54. Levie, R., Yapar, Ç., Kutyniok, G., Caire, G.: RadioUNet: Fast radio map estimation with convolutional neural networks. IEEE Transactions on Wireless Communications (TWC) **20**(6), 4001–4015 (2021)
55. Li, H., Wang, X., Zhang, Z., Zhu, W.: OOD-GNN: Out-of-distribution generalized graph neural network. IEEE Transactions on Knowledge and Data Engineering (TKDE) **35**(7), 7328–7340 (2022)
56. Li, M., Zhu, Z.: Spatial-temporal fusion graph neural networks for traffic flow forecasting. In: Proceedings of the 35th AAAI Conference on Artificial Intelligence (AAAI'21), pp. 4189–4196 (2021)
57. Li, S., Jin, X., Xuan, Y., Zhou, X., Chen, W., Wang, Y.X., Yan, X.: Enhancing the locality and breaking the memory bottleneck of transformer on time series forecasting. In: Proceedings of the 33rd International Conference on Neural Information Processing Systems (NeurIPS'19), pp. 1–13 (2019)
58. Li, Y., Fu, K., Wang, Z., Shahabi, C., Ye, J., Liu, Y.: Multi-task representation learning for travel time estimation. In: Proceedings of the 24th ACM SIGKDD Conference on Knowledge Discovery and Data Mining (KDD'18), pp. 1695–1704 (2018)
59. Li, Y., Yu, R., Shahabi, C., Liu, Y.: Diffusion convolutional recurrent neural network: Data-driven traffic forecasting. In: Proceedings of the 6th International Conference on Learning Representations (ICLR'18), pp. 1–16 (2018)
60. Li, Z., Kovachki, N.B., Azizzadenesheli, K., Bhattacharya, K., Stuart, A., Anandkumar, A., et al.: Fourier neural operator for parametric partial differential equations. In: Proceedings of the 9th International Conference on Learning Representations (ICLR'21), pp. 1–16 (2021)
61. Li, Z., Ton, J.F., Oglic, D., Sejdinovic, D.: Towards a unified analysis of random Fourier features. In: Proceedings of the 36th International Conference on Machine Learning (ICML'19), pp. 3905–3914 (2019)
62. Liang, Y., Wen, H., Nie, Y., Jiang, Y., Jin, M., Song, D., Pan, S., Wen, Q.: Foundation models for time series analysis: A tutorial and survey. In: Proceedings of the 30th ACM SIGKDD Conference on Knowledge Discovery and Data Mining (KDD'24), pp. 6555–6565 (2024)
63. Lighthill, M.J., Whitham, G.B.: On kinematic waves II. A theory of traffic flow on long crowded roads. Proceedings of the Royal Society of London **229**(1178), 317–345 (1955)
64. Lu, L., Jin, P., Pang, G., Zhang, Z., Karniadakis, G.E.: Learning nonlinear operators via DeepONet based on the universal approximation theorem of operators. Nature Machine Intelligence **3**(3), 218–229 (2021)
65. Lu, W., Wang, J., Sun, X., Chen, Y., Ji, X., Yang, Q., Xie, X.: DIVERSIFY: A general framework for time series out-of-distribution detection and generalization. IEEE Transactions on Pattern Analysis and Machine Intelligence (TPAMI) **46**(6), 4534–4550 (2024)
66. Lv, Y., Duan, Y., Kang, W., Li, Z., Wang, F.Y.: Traffic flow prediction with big data: A deep learning approach. IEEE Transactions on Intelligent Transportation Systems (ITS) **16**(2), 865–873 (2014)

67. Ma, J., Cui, P., Kuang, K., Wang, X., Zhu, W.: Disentangled graph convolutional networks. In: Proceedings of the 36th International Conference on Machine Learning (ICML'19), pp. 4212–4221 (2019)
68. Ma, X., Tao, Z., Wang, Y., Yu, H., Wang, Y.: Long short-term memory neural network for traffic speed prediction using remote microwave sensor data. Transportation Research Part C: Emerging Technologies **54**(1), 187–197 (2015)
69. Matsubara, Y., Sakurai, Y., Van Panhuis, W.G., Faloutsos, C.: FUNNEL: Automatic mining of spatially coevolving epidemics. In: Proceedings of the 20th ACM SIGKDD Conference on Knowledge Discovery and Data Mining (KDD'14), pp. 105–114 (2014)
70. May, A.D.: Traffic flow fundamentals. Prentice-Hall (1990)
71. Newell, G.F.: Nonlinear effects in the dynamics of car following. Operations Research **9**(2), 209–229 (1961)
72. Ni, D.: Traffic flow theory: Characteristics, experimental methods, and numerical techniques. Butterworth-Heinemann (2015)
73. Offenhuber, D., Ratti, C.: Decoding the City: Urbanism in the Age of Big Data. Birkhäuser (2014)
74. Pan, Z., Liang, Y., Wang, W., Yu, Y., Zheng, Y., Zhang, J.: Urban traffic prediction from spatio-temporal data using deep meta learning. In: Proceedings of the 25th ACM SIGKDD Conference on Knowledge Discovery and Data Mining (KDD'19), pp. 1720–1730 (2019)
75. Pan, Z., Wang, Z., Wang, W., Yu, Y., Zhang, J., Zheng, Y.: Matrix factorization for spatio-temporal neural networks with applications to urban flow prediction. In: Proceedings of the 28th ACM International Conference on Information and Knowledge Management (CIKM'19), pp. 2683–2691 (2019)
76. Park, H., Lee, S., Kim, S., Park, J., Jeong, J., Kim, K.M., Ha, J.W., Kim, H.J.: Metropolis-Hastings data augmentation for graph neural networks. In: Proceedings of the 35th International Conference on Neural Information Processing Systems (NeurIPS'21), pp. 19010–19020 (2021)
77. Payne, H.J.: Freeway traffic control and surveillance model. Transportation Engineering Journal of ASCE **99**(4), 767–783 (1973)
78. Pegels, C.C.: Exponential forecasting: Some new variations. Management Science **15**(5), 311–315 (1969)
79. Pipes, L.A.: An operational analysis of traffic dynamics. Journal of Applied Physics **24**(3), 274–281 (1953)
80. Pipes, L.A.: Vehicle accelerations in the hydrodynamic theory of traffic flow. Transportation Research **3**(2), 229–234 (1969)
81. Qiu, J., Jammalamadaka, S.R., Ning, N.: Multivariate Bayesian structural time series model. Journal of Machine Learning Research (JMLR) **19**(68), 1–33 (2018)
82. Qiu, X., Ren, Y., Suganthan, P.N., Amaratunga, G.A.: Empirical mode decomposition based ensemble deep learning for load demand time series forecasting. Applied Soft Computing **54**(1), 246–255 (2017)
83. Rahimi, A., Recht, B.: Random features for large-scale kernel machines. In: Proceedings of the 21st International Conference on Neural Information Processing Systems (NeurIPS'07), pp. 7328–7340 (2007)
84. Raissi, M., Perdikaris, P., Karniadakis, G.E.: Physics-informed neural networks: A deep learning framework for solving forward and inverse problems involving nonlinear partial differential equations. Journal of Computational Physics **378**(1), 686–707 (2019)
85. Ravi, S., Larochelle, H.: Optimization as a model for few-shot learning. In: Proceedings of the 5th International Conference on Learning Representations (ICLR'17), pp. 1–11 (2017)
86. Reichstein, M., Camps-Valls, G., Stevens, B., Jung, M., Denzler, J., Carvalhais, N., Prabhat, F.: Deep learning and process understanding for data-driven Earth system science. Nature **566**(7743), 195–204 (2019)
87. Reuschel, A.: Vehicle movements in the column uniformly accelerated or delayed. Oesterrich Ingr Arch **4**(1), 193–215 (1950)
88. Richards, P.I.: Shock waves on the highway. Operations Research **4**(1), 42–51 (1956)

89. Rong, Y., Huang, W., Xu, T., Huang, J.: DropEdge: Towards deep graph convolutional networks on node classification. In: Proceedings of the 8th International Conference on Learning Representations (ICLR'20), pp. 1–17 (2020)
90. Ronneberger, O., Fischer, P., Brox, T.: U-Net: Convolutional networks for biomedical image segmentation. In: Proceedings of the 18th International Conference on Medical Image Computing and Computer-Assisted Intervention, pp. 234–241 (2015)
91. Ross, P.: Traffic dynamics. Transportation Research Part B: Methodological **22**(6), 421–435 (1988)
92. Roy, J.R., Thill, J.C.: Spatial Interaction Modelling. Springer (2004)
93. Sanchez-Gonzalez, A., Godwin, J., Pfaff, T., Ying, R., Leskovec, J., Battaglia, P.: Learning to simulate complex physics with graph networks. In: Proceedings of the 37th International Conference on Machine Learning (ICML'20), pp. 8459–8468 (2020)
94. Scarselli, F., Gori, M., Tsoi, A.C., Hagenbuchner, M., Monfardini, G.: The graph neural network model. IEEE Transactions on Neural Networks and Learning Systems **20**(1), 61–80 (2008)
95. Shao, C.: Transportation Planning Theory. China Railway Publishing House (2008)
96. Shao, Z., Zhang, Z., Wei, W., Wang, F., Xu, Y., Cao, X., Jensen, C.S.: Decoupled dynamic spatial-temporal graph neural network for traffic forecasting. Proceedings of the VLDB Endowment **15**(11), 2733–2746 (2022)
97. Sheppard, E.S.: Theoretical underpinnings of the gravity hypothesis. Geographical Analysis **10**(4), 386–402 (1978)
98. Simini, F., González, M.C., Maritan, A., Barabási, A.L.: A universal model for mobility and migration patterns. Nature **484**(7392), 96–100 (2012)
99. Song, C., Lin, Y., Guo, S., Wan, H.: Spatial-temporal synchronous graph convolutional networks: A new framework for spatial-temporal network data forecasting. In: Proceedings of the 34th AAAI Conference on Artificial Intelligence (AAAI'20), pp. 914–921 (2020)
100. Song, X., Kanasugi, H., Shibasaki, R.: DeepTransport: Prediction and simulation of human mobility and transportation mode at a citywide level. In: Proceedings of the 25th International Joint Conference on Artificial Intelligence (IJCAI'16), pp. 2618–2624 (2016)
101. Soua, R., Koesdwiady, A., Karray, F.: Big-data-generated traffic flow prediction using deep learning and Dempster-Shafer theory. In: Proceedings of the 2016 International Joint Conference on Neural Networks, pp. 3195–3202 (2016)
102. Stewart, J.Q.: An inverse distance variation for certain social influences. Science **93**(2404), 89–90 (1941)
103. Stouffer, S.A.: Intervening opportunities: A theory relating mobility and distance. American Sociological Review **5**(6), 845–867 (1940)
104. Takamoto, M., Praditia, T., Leiteritz, R., MacKinlay, D., Alesiani, F., Pflüger, D., Niepert, M.: PDEBench: An extensive benchmark for scientific machine learning. Advances in Neural Information Processing Systems **35**(1), 1596–1611 (2022)
105. Toledo, T., Koutsopoulos, H.N., Ben-Akiva, M.: Integrated driving behavior modeling. Transportation Research Part C: Emerging Technologies **15**(2), 96–112 (2007)
106. Tran, A., Mathews, A., Xie, L., Ong, C.S.: Factorized Fourier neural operators. In: Proceedings of the 11th International Conference on Learning Representations (ICLR'23), pp. 1–17 (2023)
107. Treiber, M., Hennecke, A., Helbing, D.: Congested traffic states in empirical observations and microscopic simulations. Physical Review E **62**(2), 1805 (2000)
108. Tu, L., Lalwani, G., Gella, S., He, H.: An empirical study on robustness to spurious correlations using pre-trained language models. Transactions of the Association for Computational Linguistics **8**(1), 621–633 (2020)
109. Van Winsum, W.: The human element in car following models. Transportation Research Part F: Traffic Psychology and Behaviour **2**(4), 207–211 (1999)
110. Vaswani, A., Shazeer, N., Parmar, N., Uszkoreit, J., Jones, L., Gomez, A.N., Kaiser, L., Polosukhin, I.: Attention is all you need. In: Proceedings of the 31st International Conference on Neural Information Processing Systems (NeurIPS'17), pp. 5998–6008 (2017)

111. Velickovic, P., Cucurull, G., Casanova, A., Romero, A., Liò, P., Bengio, Y.: Graph attention networks. In: Proceedings of the 6th International Conference on Learning Representations (ICLR'18), pp. 1–12 (2018)
112. Wang, B., Ma, J., Wang, P., Wang, X., Zhang, Y., Zhou, Z., Wang, Y.: STONE: A spatio-temporal OOD learning framework kills both spatial and temporal shifts. In: Proceedings of the 30th ACM SIGKDD Conference on Knowledge Discovery and Data Mining (KDD'24), pp. 2948–2959 (2024)
113. Wang, D., Cao, W., Li, J., Ye, J.: DeepSD: Supply-demand prediction for online car-hailing services using deep neural networks. In: Proceedings of the 2017 IEEE 33rd International Conference on Data Engineering (ICDE'17), pp. 243–254 (2017)
114. Wang, J., Gu, Q., Wu, J., Liu, G., Xiong, Z.: Traffic speed prediction and congestion source exploration: A deep learning method. In: Proceedings of the 2016 IEEE International Conference on Data Mining (ICDM'16), pp. 499–508 (2016)
115. Wang, J., Wang, Z., Li, J., Wu, J.: Multilevel wavelet decomposition network for interpretable time series analysis. In: Proceedings of the 24th ACM SIGKDD Conference on Knowledge Discovery and Data Mining (KDD'18), pp. 2437–2446 (2018)
116. Wang, Y., Zou, J., Lin, J., Ling, Q., Pan, Y., Yao, T., Mei, T.: Out-of-distribution detection via conditional kernel independence model. In: Proceedings of the 36th International Conference on Neural Information Processing Systems (NeurIPS'22), pp. 36411–36425 (2022)
117. Wei, H., Wang, Y., Wo, T., Liu, Y., Xu, J.: ZEST: A hybrid model on predicting passenger demand for chauffeured car service. In: Proceedings of the 25th ACM International Conference on Information and Knowledge Management (CIKM'16), pp. 2203–2208 (2016)
118. Wen, G., Li, Z., Azizzadenesheli, K., Anandkumar, A., Benson, S.M.: U-FNO—An enhanced Fourier neural operator-based deep-learning model for multiphase flow. Advances in Water Resources **163**(1), 104180 (2022)
119. Wilson, A.G.: A statistical theory of spatial distribution models. Transportation Research **1**(3), 253–269 (1967)
120. Woo, G., Liu, C., Sahoo, D., Kumar, A., Hoi, S.: CoST: Contrastive learning of disentangled seasonal-trend representations for time series forecasting. In: Proceedings of the 10th International Conference on Learning Representations (ICLR'22), pp. 1–18 (2022)
121. Wu, Y.X., Wang, X., Zhang, A., He, X., Chua, T.S.: Discovering invariant rationales for graph neural networks. In: Proceedings of the 10th International Conference on Learning Representations (ICLR'22), pp. 1–22 (2022)
122. Wu, Z., Pan, S., Long, G., Jiang, J., Chang, X., Zhang, C.: Connecting the dots: Multivariate time series forecasting with graph neural networks. In: Proceedings of the 26th ACM SIGKDD Conference on Knowledge Discovery and Data Mining (KDD'20), pp. 753–763 (2020)
123. Wu, Z., Pan, S., Long, G., Jiang, J., Zhang, C.: Graph WaveNet for deep spatial-temporal graph modeling. In: Proceedings of the 28th International Joint Conference on Artificial Intelligence (IJCAI'19), pp. 1907–1913 (2019)
124. Xia, Y., Liang, Y., Wen, H., Liu, X., Wang, K., Zhou, Z., Zimmermann, R.: Deciphering spatio-temporal graph forecasting: A causal lens and treatment. In: Proceedings of the 37th International Conference on Neural Information Processing Systems (NeurIPS'23), pp. 37068–37088 (2023)
125. Xu, J., Rahmatizadeh, R., Bölöni, L., Turgut, D.: Real-time prediction of taxi demand using recurrent neural networks. IEEE Transactions on Intelligent Transportation Systems (ITS) **19**(8), 2572–2581 (2017)
126. Yan, X.: A modified calibration method for gravity model. Transportation and Computer **21**(4), 93–95 (2003)
127. Yan, X.Y., Han, X.P., Wang, B.H., et al.: Diversity of individual mobility patterns and emergence of aggregated scaling laws. Scientific Reports **3**(1), 2678 (2013)
128. Yang, C., Wu, Q., Wen, Q., Zhou, Z., Sun, L., Yan, J.: Towards out-of-distribution sequential event prediction: A causal treatment. In: Proceedings of the 36th International Conference on Neural Information Processing Systems (NeurIPS'22), pp. 22656–22670 (2022)

129. Yang, Q., Koutsopoulos, H.N.: A microscopic traffic simulator for evaluation of dynamic traffic management systems. Transportation Research Part C: Emerging Technologies **4**(3), 113–129 (1996)
130. Yao, H., Liu, Y., Wei, Y., Tang, X., Li, Z.: Learning from multiple cities: A meta-learning approach for spatial-temporal prediction. In: Proceedings of the Web Conference 2019 (WWW'19), pp. 2181–2191 (2019)
131. Yao, H., Tang, X., Wei, H., Zheng, G., Li, Z.: Revisiting spatial-temporal similarity: A deep learning framework for traffic prediction. In: Proceedings of the 33rd AAAI Conference on Artificial Intelligence (AAAI'19), pp. 5668–5675 (2019)
132. Ye, X., Fang, S., Sun, F., Zhang, C., Xiang, S.: Meta graph transformer: A novel framework for spatial-temporal traffic prediction. Neurocomputing **491**(1), 544–563 (2022)
133. Yehudai, G., Fetaya, E., Meirom, E., Chechik, G., Maron, H.: From local structures to size generalization in graph neural networks. In: Proceedings of the 38th International Conference on Machine Learning (ICML'21), pp. 11975–11986 (2021)
134. Yu, B., Yin, H., Zhu, Z.: Spatio-temporal graph convolutional networks: A deep learning framework for traffic forecasting. In: Proceedings of the 27th International Joint Conference on Artificial Intelligence (IJCAI'18), pp. 3634–3640 (2018)
135. Yu, F., Koltun, V.: Multi-scale context aggregation by dilated convolutions. In: Proceedings of the 4th International Conference on Learning Representations (ICLR'16), pp. 1–13 (2016)
136. Yu, R., Li, Y., Shahabi, C., Demiryurek, U., Liu, Y.: Deep learning: A generic approach for extreme condition traffic forecasting. In: Proceedings of the 2017 SIAM International Conference on Data Mining (SDM'17), pp. 777–785 (2017)
137. Yuan, Z., Zhou, X., Yang, T.: Hetero-ConvLSTM: A deep learning approach to traffic accident prediction on heterogeneous spatio-temporal data. In: Proceedings of the 24th ACM SIGKDD Conference on Knowledge Discovery and Data Mining (KDD'18), pp. 984–992 (2018)
138. Zhang, J., Zheng, Y., Qi, D.: Deep spatio-temporal residual networks for citywide crowd flows prediction. In: Proceedings of the 31st AAAI Conference on Artificial Intelligence (AAAI'17), pp. 1655–1661 (2017)
139. Zhang, J., Zheng, Y., Qi, D., Li, R., Yi, X.: DNN-based prediction model for spatio-temporal data. In: Proceedings of the 24th ACM SIGSPATIAL International Conference on Advances in Geographic Information Systems, pp. 1–4 (2016)
140. Zhang, X., Cui, P., Xu, R., Zhou, L., He, Y., Shen, Z.: Deep stable learning for out-of-distribution generalization. In: Proceedings of the 2021 IEEE Conference on Computer Vision and Pattern Recognition (CVPR'21), pp. 5372–5382 (2021)
141. Zhang, X., Huang, C., Xu, Y., Xia, L., Dai, P., Bo, L., Zhang, J., Zheng, Y.: Traffic flow forecasting with spatial-temporal graph diffusion network. In: Proceedings of the 35th AAAI Conference on Artificial Intelligence (AAAI'21), pp. 15008–15015 (2021)
142. Zhang, Y., Pal, S., Coates, M., Ustebay, D.: Bayesian graph convolutional neural networks for semi-supervised classification. In: Proceedings of the 33rd AAAI Conference on Artificial Intelligence (AAAI'19), pp. 5829–5836 (2019)
143. Zhang, Y., Zhang, Y., Haghani, A.: A hybrid short-term traffic flow forecasting method based on spectral analysis and statistical volatility model. Transportation Research Part C: Emerging Technologies **43**(1), 65–78 (2014)
144. Zhao, L., Song, Y., Zhang, C., Liu, Y., Wang, P., Lin, T., Deng, M., Li, H.: T-GCN: A temporal graph convolutional network for traffic prediction. IEEE Transactions on Intelligent Transportation Systems (ITS) **21**(9), 3848–3858 (2019)
145. Zhao, Z., Chen, W., Wu, X., Chen, P.C., Liu, J.: LSTM network: A deep learning approach for short-term traffic forecast. IET Intelligent Transport Systems **11**(2), 68–75 (2017)
146. Zhao, Z., Ding, X., Prakash, B.A.: PINNsFormer: A transformer-based framework for physics-informed neural networks. In: Proceedings of the 12th International Conference on Learning Representations (ICLR'24), pp. 1–17 (2024)
147. Zhou, G., Mou, N., Fan, Y., Pi, Q., Bian, W., Zhou, C., Zhu, X., Gai, K.: Deep interest evolution network for click-through rate prediction. In: Proceedings of the 33rd AAAI Conference on Artificial Intelligence (AAAI'19), pp. 5941–5948 (2019)

148. Zhou, H., Zhang, S., Peng, J., Zhang, S., Li, J., Xiong, H., Zhang, W.: Informer: Beyond efficient transformer for long sequence time-series forecasting. In: Proceedings of the 35th AAAI Conference on Artificial Intelligence (AAAI'21), pp. 11106–11115 (2021)
149. Zhou, W., Wang, W., Hua, X., Zhang, Y.: Real-time traffic flow forecasting via a novel method combining periodic-trend decomposition. Sustainability **12**(15), 5891 (2020)
150. Zhou, Z., Huang, Q., Yang, K., Wang, K., Wang, X., Zhang, Y., Liang, Y., Wang, Y.: Maintaining the status quo: Capturing invariant relations for OOD spatiotemporal learning. In: Proceedings of the 29th ACM SIGKDD Conference on Knowledge Discovery and Data Mining (KDD'23), pp. 3603–3614 (2023)

Chapter 3
Potential Energy Field Model

Abstract Region-level traffic prediction is a key technology in intelligent transportation systems, providing high spatiotemporal resolution information on traffic flow and congestion. However, existing deep learning prediction models lack modeling of the underlying mechanisms of crowd movement. To address this, this chapter introduces a gravity-inspired, discrete-time spatiotemporal potential field model that can accurately predict future regional traffic flow while explaining its evolution.

3.1 Conceptual Foundations

Region-level traffic prediction is a fundamental problem in intelligent transportation systems (ITS). It provides high spatio-temporal resolution forecasts of traffic volume, demand, and congestion, thereby supporting precise decision-making for urban management and intelligent mobility.

On the management side, accurate region-level forecasts enable signal timing optimization to reduce delays and improve throughput, and offer quantitative evidence for traffic control and restriction policies. They also assist emergency response by predicting the impact of incidents or extreme weather and informing rescue and evacuation planning. For intelligent mobility services, region-level prediction supplies real-time demand information for vehicle dispatching, helping drivers select profitable destinations, improving user experience, and increasing both driver income and platform satisfaction. Moreover, by reducing empty mileage and unnecessary cruising, accurate forecasts contribute to lower carbon emissions and more sustainable urban development.

The grid-based method partitions the urban space into regular cells, each representing a region. Owing to its flexibility and generality, this representation has been widely adopted in traffic prediction, environmental monitoring, and urban computing. The central idea is to discretize urban space so that complex geospatial data can be represented and processed within a unified, structured grid system. Concretely, the city is divided into fixed-size rectangular cells according to latitude-longitude, road layout, or administrative boundaries; each cell aggregates region-level spatio-temporal features such as traffic volume or crowd density. Partitioning the city into

J. Wang and J. Ji, *Mechanism-Driven Explainable Urban Spatio-Temporal Prediction*, Big Data Management, https://doi.org/10.1007/978-981-92-0662-9_3

a regular grid[1] flexibly adapts to complex urban structures and can be applied to roads, open spaces, pedestrian flows, and shared-mobility hotspots without relying on a particular road-network topology. The regular structure also integrates well with deep models (e.g., CNNs and RNNs), which leverage image- and sequence-modeling techniques to learn spatial correlations and temporal dependencies. Accordingly, this chapter studies inter-regional traffic flow on grid-based data.

A large body of work has explored learning-based models for region-level traffic forecasting. Early studies mainly relied on shallow statistical models that treat traffic as a univariate or multivariate time series, employing ARIMA and Kalman filtering to capture temporal dynamics [24]. More recent approaches typically learn temporal and spatial dependencies with deep models, for example RNNs and their variants for dynamic temporal dependencies [5, 29] and CNNs for spatial correlations in gridded traffic data [31, 32].

Despite this progress, three key limitations remain.

- *First*, most region-level predictors overlook the underlying mechanisms of human mobility. They are primarily data-driven, learning spatio-temporal patterns from historical flows without explicitly modeling the generative process. The absence of such a mechanism weakens interpretability and generalization. In practice, traffic evolution is driven not only by past volumes but also by behavioral regularities and physical constraints; ignoring these factors risks unstable or inaccurate predictions.
- *Second*, insufficient attention to *local* spatial correlations hinders precise modeling of inter-regional interactions. Some methods model each region independently and ignore spatial relationships; others adopt a global scheme that assumes homogeneous spatial patterns across all regions. Both strategies fail to capture strong local dependencies that are salient in traffic: regions of the same functional type often share similar internal flow patterns, whereas flows across heterogeneous regions (e.g., commercial vs. residential vs. industrial) differ markedly. A purely global model may underfit these differences and degrade accuracy.
- *Third*, inductive biases in common deep architectures limit their ability to capture the *directionality* of flows. For instance, CNNs extract local spatial features but their kernels do not encode directed flow (inflow/outflow), and many GNNs are designed for undirected graphs, whereas traffic is inherently directed: flows propagate from one region to another along specific directions. Standard temporal modules also tend to employ a single mechanism, which is ill-suited to the multiple temporal patterns present in traffic.

To address these gaps, we propose a discrete-time, gravity-inspired *Spatio-Temporal Potential Energy Field* (PEF) model [22]. The model adopts a gravity formulation to quantify inter-regional attraction and, together with historical traffic flows, defines an algorithm to extract discrete spatio-temporal potential energy fields that drive human mobility. By injecting physical priors derived from gravity-based mobility theory, the PEF provides physically meaningful spatio-temporal features for

[1] For readability, we do not distinguish between the terms "grid" and "raster" in this chapter.

downstream deep forecasting, enhancing generalization while offering mechanism-based explanations via potential-field dynamics. This, in turn, improves interpretability and trustworthiness. For brevity, we henceforth refer to the "spatio-temporal potential energy field" as the *potential energy field* (PEF) and denote the PEF-based spatio-temporal predictor as *STPEF*.

First, to more faithfully capture the intrinsic mechanisms governing urban traffic flows, we develop a gravity-theoretic PEF model together with a discrete-time algorithm for extracting region-level PEFs. Although gravity theory has long been employed to describe and predict human mobility [2, 15], classical gravity models are typically parameter-sparse and thus lack the flexibility needed to characterize complex traffic phenomena. We therefore introduce a hybrid mechanism, termed *Potential Energy Fields* (PEFs), that is jointly grounded in gravity theory and real spatio-temporal observations. By analogy with water moving along gradients in a gravitational field [6, 16], the PEF hypothesis assumes that human movement tends to occur from high- to low-potential locations. To recover latent PEFs from observed traffic flows, we design a decomposition algorithm that partitions the grid-based traffic flow graph into a collection of *polytree* subgraphs. Each polytree induces a PEF in which every node is assigned a unique relative potential and flows are allowed only from higher- to lower-potential nodes. Under this formulation, the observed traffic graph can be interpreted as the superposition of multiple simple PEFs. Furthermore, we propose a novel *continuous boundary scaling* procedure and theoretically prove that the resulting algorithm can decompose any grid-based traffic flow graph into PEFs.

Second, to remedy the neglect of local spatial correlations, we propose a community partitioning algorithm that is *global-flow-aware* in both time and space, enabling the identification of highly correlated subregions. Community detection aims to partition a network into communities with dense intra-community connections and sparse inter-community links [13]. By injecting global traffic information over long historical horizons into this process, the PEF model can more precisely capture local flow patterns, and the resulting partitions reflect long-term, stable mobility structures. Beyond intra-community PEF extraction, we further generalize the framework to inter-community flows, thereby enabling hierarchical modeling of the spatial topology in spatio-temporal data. Notably, the proposed algorithm does not require gridded inputs and extends naturally to non-Euclidean structures, paving the way for PEF extraction on realistic road networks.

Third, to capture the complex temporal and spatial dependencies in urban traffic, we develop a correlation-adaptive deep forecasting architecture. Temporal dependencies consist of short-term correlations and long-term periodicities with distinct properties [26]: long-term cycles generalize well and provide stable trends, whereas short-term correlations dominate at small time lags and are critical for modeling recent dynamics. Accordingly, we employ different neural structures for the two regimes: a gated recurrent unit (GRU) *without* attention for short-term dynamics, which leverages recurrence to exploit recent observations, and a dilated causal convolutional network (DCN) for long-term periodic patterns. This dual-temporal design jointly learns short-term fluctuations and long-term regularities, thereby improving both accuracy and generalization. On the spatial side, we inject the directed, weighted polytree structure of the PEF into a graph attention network, yielding a *directed, weight-aware attention* mechanism that more faithfully captures the dynamic, directional spatial dependencies encoded by the PEF.

3.2 Core Concepts and Modeling Objectives

Before presenting the proposed model, we first formalize the region-level traffic flow prediction problem considered in this chapter. Given the map of a target city, we discretize the urban space into a raster grid and treat each grid cell as the basic computational unit, upon which we construct a mechanism-inspired model and a deep learning predictor for regional traffic flows.

Definition 3.1 (*Spatial Grid*) A *spatial grid* $\boldsymbol{S} \in \mathbb{R}^{I \times J}$ represents a regular partition of the urban space, where I and J are the numbers of rows and columns, respectively. The study area is divided into $I \times J$ grid cells, each corresponding to a specific region and treated as the minimal prediction unit.

Figure 3.1a shows an example spatial grid with 4×4 regions. For each cell, flows are categorized into eight types, corresponding to inflow and outflow with its four

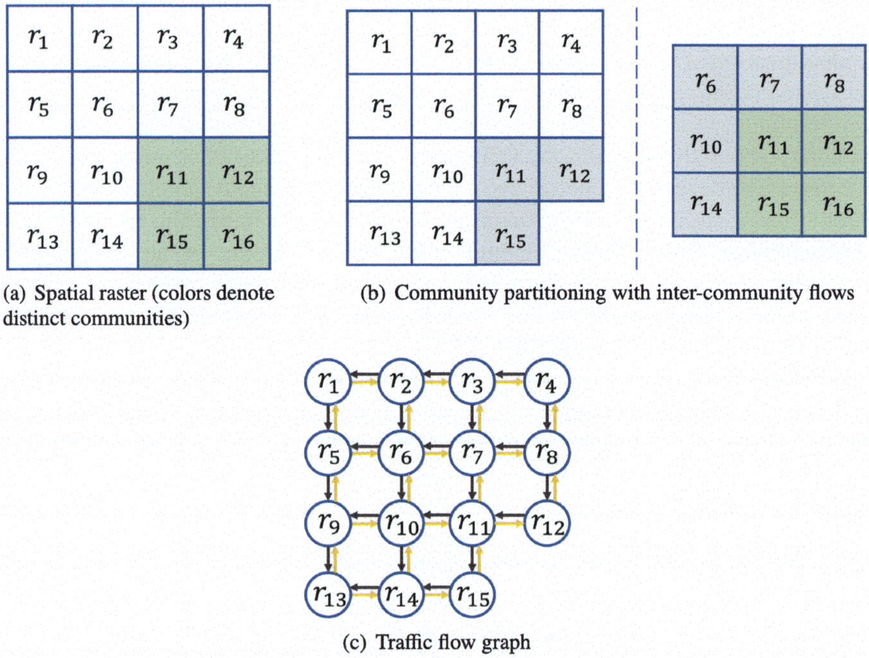

Fig. 3.1 Illustrations of the spatial raster and traffic flow graph

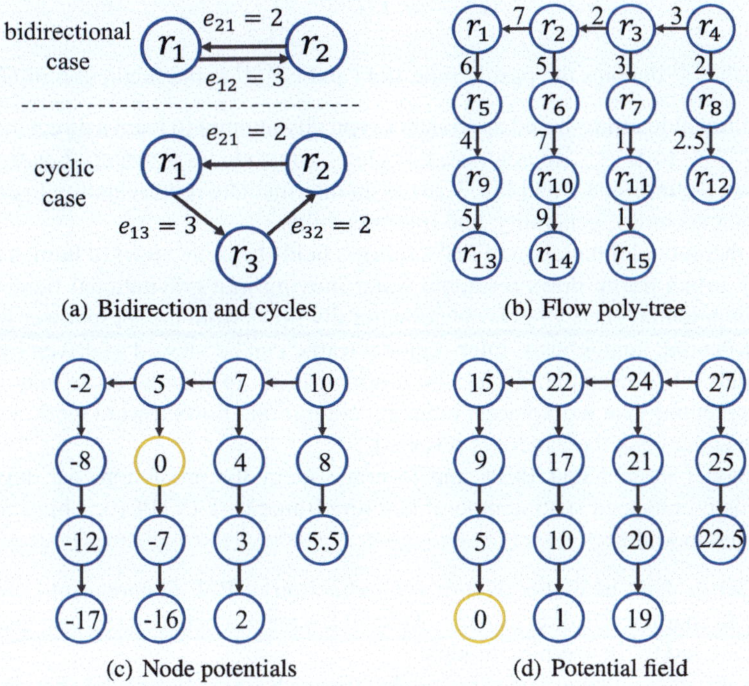

Fig. 3.2 Concepts related to grid-based potential energy fields

neighbors (up, right, down, and left). Based on the grid structure, we construct a traffic flow graph to model inter-regional movements as follows.

Definition 3.2 (*Regional Traffic Flow Graph*) A *regional traffic flow graph* $G(V_G, E_G)$ is a weighted, directed graph. The node set V_G indexes all regions (grid cells) and the edge set E_G represents directed traffic flows between neighboring regions. Concretely, $v_i \in V_G$ denotes a grid cell, and the directed flow from v_i to v_j is denoted $e_{ij} \in E_G$.

For brevity, we refer to the regional traffic flow graph simply as the *flow graph*.

Definition 3.3 (*Sequence of Regional Traffic Flow Graphs*) Time is discretized into regular slots. For each slot t, there is a corresponding flow graph $G^{(t)}$ over the spatial grid. The temporal evolution of regional flows is thus represented by a sequence $\mathcal{G}^{(1:t)} = \left(G^{(1)}, G^{(2)}, \ldots, G^{(t)}\right)$.

When no ambiguity arises, we call $\mathcal{G}^{(1:t)}$ the *flow graph sequence*. We now formalize the prediction task.

Definition 3.4 (*Regional Traffic Flow Prediction*) Given the historical flow graph sequence from time $t - T$ to t, the goal is to learn a function f that predicts the next-time-step flow state:

$$\hat{G}^{(t+1)} = f\left(\mathcal{G}^{(t-T:t)}\right), \tag{3.1}$$

where $\mathcal{G}^{(t-T:t)}$ denotes the past T time slots and $\hat{G}^{(t+1)}$ is the prediction of $G^{(t+1)}$.

Traditional learning-based approaches typically attempt to learn a direct mapping from $\mathcal{G}^{(t-T:t)}$ to $G^{(t+1)}$. In contrast, this chapter introduces a *potential energy field* (PEF; see Definition 3.8 and Fig. 3.2d) as an intermediate representation to enhance prediction accuracy, generality, and interpretability.

The theoretical foundation of PEFs follows field-theoretic views of human mobility [11]: urban traffic flows resemble water moving in a gravitational field, driven by latent potential energy fields. Water naturally flows from higher to lower gravitational potential; analogously, inter-regional traffic can be viewed as driven by latent "potentials." For example, during the morning peak, residential areas may exhibit higher potential than workplaces, inducing commuting flows toward work, whereas in the evening peak the pattern is reversed.

To model these latent PEFs and leverage them for prediction, we develop a mechanism-enhanced spatiotemporal learning framework (STPEF), which consists of two key modules:

- *PEF extraction*: heuristic, data-driven extraction of PEF features from observed flows to characterize directional movement tendencies between regions; these features are fed to the learning module.
- *Data-driven predictor*: a deep model that forecasts future flows while conditioning on the extracted PEF features to improve expressiveness and accuracy.

This design preserves the predictive strength of deep models while providing mechanistic interpretability via PEFs, revealing intrinsic drivers of traffic evolution.

3.3 Spatio-Temporal Potential Energy Field Extraction

In the PEF extraction module over regional flow graphs, we introduce *directional decomposition* and *poly-tree decomposition* to convert a flow graph into multiple PEFs. We assume urban flow is driven by latent PEFs, akin to gravity-driven water flow, but traffic exhibits more complex dynamics than physical fluids. Two key differences are:

- *Directionality*: Water flows in a single direction along the potential gradient, whereas traffic between two locations can be bidirectional. If r_1 and r_2 are adjacent, water can only move $r_1 \to r_2$ or $r_2 \to r_1$, but traffic may exhibit both directions simultaneously.
- *Cycles*: Gravity-driven water flow does not form cycles (e.g. $r_1 \to r_2 \to r_3 \to r_1$), yet cyclic flows are prevalent in traffic graphs.

These differences complicate potential assignment. With bidirectional edges or cycles, a unique scalar potential per node may not exist. For example, if both $r_1 \to r_2$

and $r_2 \rightarrow r_1$ carry positive flow, each direction would imply a contradictory potential ordering; similarly, a cycle $r_1 \rightarrow r_2 \rightarrow r_3 \rightarrow r_1$ yields inconsistent inequalities.

As an illustration, Fig. 3.2a shows a bidirectional case where flow from r_1 to r_2 is 3 and from r_2 to r_1 is 2. Setting r_1 as the zero-potential reference, $r_1 \rightarrow r_2$ suggests $p_2 = -3$, whereas $r_2 \rightarrow r_1$ suggests $p_2 = 2$, a contradiction. The same figure also shows a cyclic case with $e_{21} = 2, e_{13} = 3$, and $e_{32} = 2$. Setting r_1 to zero, path $r_2 \rightarrow r_1$ gives $p_2 = 2$, while path $r_1 \rightarrow r_3 \rightarrow r_2$ gives $p_2 = -5$.

To address these issues, we decompose the observed flow graph into multiple PEFs. Each PEF individually satisfies unidirectionality and acyclicity; the complex behavior of the original graph is the superposition of several simple PEFs. Our PEF extraction pipeline comprises three steps:

(i) *Community partitioning*: analyze the graph topology to identify highly correlated subregions for localized modeling.
(ii) *Directional decomposition*: resolve bidirectional edges by splitting flows into independent one-way components consistent with monotone potentials.
(iii) *Poly-tree decomposition*: eliminate cycles by decomposing the graph into multiple directed, acyclic tree-structured subgraphs (poly-trees), ensuring well-defined potentials within each PEF.

This pipeline captures latent driving mechanisms of traffic and provides physically intuitive features for forecasting.

3.3.1 Community Partitioning on the Spatial Grid

As the first step, *community partitioning* (CP) divides the study area into spatially compact communities (see the green region in Fig. 3.1a). This reduces graph complexity, improves modeling efficiency, and enhances interpretability of the resulting PEFs. It also helps capture local flow characteristics so that each PEF better describes intra-community movement.

The objective is to maximize intra-community flow while minimizing inter-community interactions. We first aggregate flows over the time window $[t - T, t]$ to form

$$\widetilde{G} = \sum_{i=t-T}^{t} G^{(i)}, \tag{3.2}$$

where $\widetilde{G}$ captures the overall flow pattern in the window. We then partition $\widetilde{G}$ into connected subgraphs (communities):

$$\{G_c \mid c \in C\} = \mathrm{CP}(\widetilde{G}), \tag{3.3}$$

where CP denotes the partitioning operator, C is the set of communities, and G_c is the subgraph induced by community c.

We adopt an improved Girvan-Newman (GN) algorithm [13], which iteratively removes edges until the graph is split into a desired number of subgraphs. Edge removal is guided by *edge betweenness*, defined as the number of shortest paths that pass through a given edge. For weighted flow graphs, we treat an edge with weight n as n parallel unweighted edges so that edge betweenness more faithfully reflects flow magnitude.

To incorporate cross-community flows into PEFs, we apply a *boundary expansion* step. For each community, we merge neighboring cells from adjacent communities along the community boundary into the current community (gray cells in Fig. 3.1b). This creates overlapping boundary regions, ensuring that inter-community flows are preserved and modeled as flows between the expanded boundary cells and the original community cells.

For stability during modeling and training, we use a fixed community structure shared across all sequences, reflecting long-term patterns in urban mobility. Accordingly, we compute communities from sufficiently long historical data. The same framework can be extended to dynamic clustering to adapt to time-varying flow patterns [3].

3.3.2 Directional Decomposition of the Flow Graph

Directional decomposition addresses bidirectional edges to ensure definable, monotone potentials. Specifically, each community flow graph is split into two subgraphs that contain only one-way edges.

Given the aggregated flow graph $\widetilde{G}$, we define the bottom-to-top edge set $E_{\uparrow}$ and the top-to-bottom set $E_{\downarrow}$. The direction with larger total flow is deemed *major*, the other *minor*. Similarly, we define left-to-right $E_{\rightarrow}$ and right-to-left $E_{\leftarrow}$ and determine their major/minor roles by total flow.

As illustrated in Fig. 3.1c, each flow graph is decomposed into:

- *Major-flow graph (black edges)*: edges along the dominant direction;
- *Minor-flow graph (yellow edges)*: edges along the weaker direction.

In both subgraphs, adjacent nodes are connected only by one-way edges, removing bidirectionality and restoring monotone potentials, which facilitates subsequent PEF construction.

3.3.3 Poly-Tree Decomposition

Polytree decomposition (PTD) removes cycles from both the major and minor subgraphs. A *polytree* is a directed acyclic graph (DAG) whose underlying undirected structure forms a tree [14]. By decomposing the flow graph into a collection of polytrees, PTD eliminates directed cycles and yields well-defined node potentials. In what follows, we formalize the polytree decomposition problem, analyze its solvability, present an efficient decomposition algorithm, and describe how to generate potential fields from the resulting polytrees.

3.3.3.1 Problem Definition

Definition 3.5 (*Flow Poly-Tree*) A *flow poly-tree* $T(V_T, E_T)$ is a directed spanning tree of the flow grid graph $G(V_G, E_G)$. That is, it covers all nodes ($V_T = V_G$) but its edges form a subset of the original edges ($E_T \subset E_G$), ensuring acyclicity.

For example, Fig. 3.2b shows a poly-tree corresponding to the major-flow graph (black) in Fig. 3.1c. It shares the same node set but uses a subset of edges to maintain a tree structure.

Definition 3.6 (*Poly-Tree Decomposition*) *Poly-tree decomposition* seeks to partition a flow graph G_g into k edge-disjoint flow poly-trees $T_1, T_2, \ldots, T_k$ such that $E_{T_i} \cap E_{T_j} = \emptyset$ for $i \neq j$ and $\bigcup_{i=1}^{k} E_{T_i} = E_{G_g}$.

The strict edge-disjointness may render the problem infeasible. We therefore use an *edge splitting* technique: each directed edge can be split into multiple parallel directed edges whose total flow equals the original flow; the split edges are assigned to different poly-trees, ensuring disjointness. This guarantees solvability in theory (see below).

3.3.3.2 Theoretical Analysis

We begin with a foundational lemma:

Lemma 3.1 *A finite graph G admits k edge-disjoint spanning trees if and only if* $\Delta_G(V_G) = 0$ *and, for every nonempty subset* $X \subseteq V_G$, $\Delta_G(X) \geq 0$, *where*

$$\Delta_G(X) = k\big(|X| - 1\big) - |E_X|, \tag{3.4}$$

and $|\cdot|$ *denotes set cardinality.*

We refer the reader to Nash-Williams for a detailed proof [12]. Building on this lemma, we propose a *continuous boundary scaling* method, which progressively relaxes the edge-count bounds so that the lemma's conditions are satisfied, thereby establishing that the PTD problem is always solvable for $k = 2$.

Theorem 3.1 *When* $k = 2$, *the PTD problem in Definition 3.6 always has a feasible solution.*

Proof By Lemma 3.1, a grid-based flow graph G must satisfy: (i) for any nonempty $X \subseteq V_G$, $\Delta_G(X) \geq 0$; and (ii) $\Delta_G(V_G) = 0$. We examine both.
Condition (i):

From the lemma,

$$k \geq p(X) = \frac{|E_X|}{|X| - 1}. \tag{3.5}$$

Determining k is nontrivial because $|E_X|$ may vary widely. We therefore bound $p(X)$ via two relaxations. First, the number of edges in a grid subgraph satisfies

$$|E_X| \leq |E_X^u| = 2|X| - \text{ceil}\left(2\sqrt{|X|}\right), \tag{3.6}$$

yielding

$$p(X) \leq \frac{2|X| - \text{ceil}\left(2\sqrt{|X|}\right)}{|X| - 1}. \tag{3.7}$$

Further relax using $\sqrt{|X|} \geq 1$ so that

$$|E_X^u| \leq 2|X| - 2, \tag{3.8}$$

and obtain

$$p(X) \leq \frac{2|X| - 2}{|X| - 1} = 2. \tag{3.9}$$

Hence $k \geq 2$ suffices for (i).
Condition (ii):
For the whole graph,

$$k = \frac{|E_G|}{(|V_G| - 1)}. \tag{3.10}$$

Using Eqs. (3.6) and (3.8),

$$|E_G| \leq 2|V_G| - 2 \quad \Rightarrow \quad k \leq 2. \tag{3.11}$$

Combining (i) and (ii), we conclude that $k = 2$ ensures feasibility. If $|E_G| < 2(|V_G| - 1)$, we employ edge splitting to augment parallel edges until $\Delta_G(V_G) = 0$ holds, preserving feasibility. □

The continuous boundary scaling idea—bounding edge counts and scaling constraints to satisfy solvability—extends beyond PTD and may be useful for other grid-graph analyses.

3.3.3.3 Algorithm Design

Guided by Theorem 3.1, PTD decomposes each (major/minor) subgraph into two *comb-like* poly-trees (see Fig. 3.2b). For each poly-tree, we first determine the backbone and tooth directions, then alternate expansions along these directions. Algo-

Algorithm 1: Poly-Tree Decomposition (PTD)

```
   Input: Flow graph $G(V_G, E_G)$
   Output: Two flow poly-trees $\{T_1, T_2\}$
1  Set the backbone and tooth directions for $T_1$
2  Set the backbone and tooth directions for $T_2$ orthogonal to those of $T_1$
3  for $i = 1$ to 2 do
4      $T_i \leftarrow \emptyset$
5      Set the origin of the max-flow edge as $b_i$
6      repeat
7          Extend $b_i$ bidirectionally along the backbone to form $G^b$
8          Add $G^b$ to $T_i$
9          Extend all nodes in $V_{G^b}$ bidirectionally along the tooth direction to form $G^t$
10         Add $G^t$ to $T_i$
11         if $V_{T_i} \neq V_G$ then
12             for $v^t \in V_{G^t}$ do
13                 Extend along the backbone to reach node $m$
14                 if adding $m$ does not create a cycle in $T_i$ then
15                     Add $m$ to $V_{G^b}$
16                     $b_i \leftarrow m$
17                     break
18                 end
19             end
20         end
21     until no further growth is possible
22 end
23 for $e \in E_{T_1}$ do
24     if $e \in E_{T_2}$ then
25         Halve the flow of $e$ in both $T_1$ and $T_2$
26     end
27 end
28 return $\{T_1, T_2\}$
```

rithm 1 provides pseudocode. Figure 3.3 visualizes the steps on a toy example. An animated demo based on the Beijing Taxi dataset is available online.[2]

Figure 3.3a: Lines 1–2 set backbone/tooth directions. For T_1, choose the direction of the maximum-flow edge as the backbone and its orthogonal as the tooth; T_2 uses directions orthogonal to T_1. In the example, since $e_{9,10}$ has the largest flow, the backbone of T_1 is $r_9 \rightarrow r_{10}$ (west$\rightarrow$east).

Figure 3.3b: Lines 5–8 initialize b_i as the origin of the max-flow edge (here r_9) and extend along the backbone to form G^b by recursively adding adjacent backbone neighbors, yielding $V_{G^b} = \{r_9, r_{10}, r_{11}, r_{12}\}$.

Figure 3.3c: Lines 9–10 extend all nodes in G^b along the tooth direction to form G^t by recursively adding tooth-direction neighbors. In the example, vertical expansions of each backbone node produce $V_{G^t} = \{r_5, r_6, r_7, r_8, r_4, r_2\}$.

[2] Animated demos of the decomposition: https://youtu.be/kAjygrBwhn4.

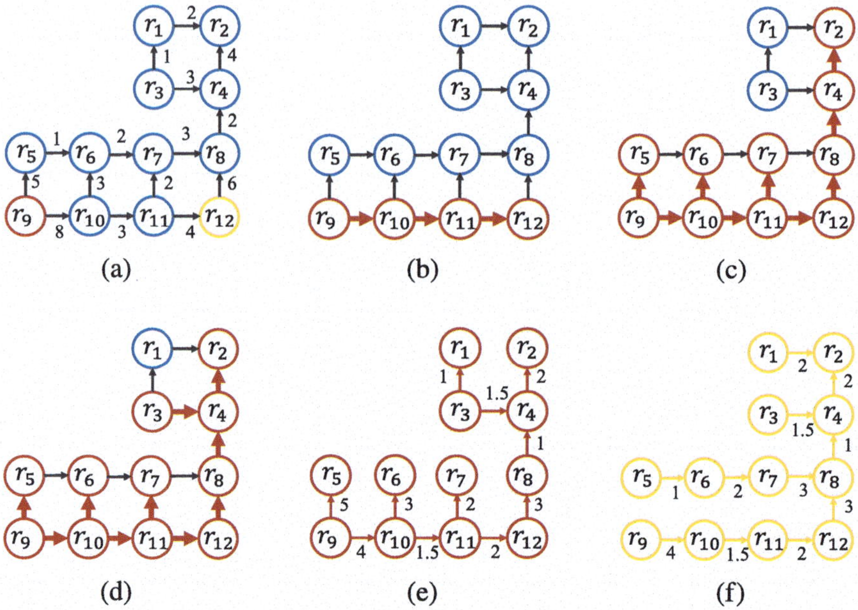

Fig. 3.3 Execution of the PTD algorithm

Figure 3.3d: Lines 11–17 attempt to expand G^b with candidates from V_{G^t} along the backbone while preventing cycles. If $V_{G^b} \cup V_{G^t} = V_G$, stop; otherwise, iterate candidates and add the first backbone neighbor that does not create a cycle (here, $r_4 \rightarrow r_3$), updating V_{G^b}.

Figure 3.3e: After adding r_3 to V_{G^b}, lines 9-10 extend its tooth neighbors (here to r_1). When all nodes are covered, T_1 is complete.

Figure 3.3f: Repeat lines 3–18 to construct T_2 (yellow). If an edge appears in both T_1 and T_2, lines 19–21 halve its flow in both to conserve totals while meeting disjointness.

Although the default start node is chosen as the origin of the maximum-flow edge, selecting a different start node merely produces a different, yet still valid, pair of polytrees. Since PTD serves as a representation of the input graph, downstream deep models—with their strong generalization ability—remain effective as long as the decomposition is edge-disjoint and exactly covers all original edges.

3.3.3.4 Potential Field Generation

Given a flow poly-tree, node potentials and the associated field are defined as follows.

Definition 3.7 (*Node Potential*) For a poly-tree, the potential of node v_i is

$$p_{v_i} = p_{v_j} + e_{ij}, \tag{3.12}$$

where e_{ij} is the flow from v_i to v_j. By selecting a reference node with zero potential, all node potentials are determined up to a constant shift.

Definition 3.8 (*Potential Energy Field*) A *potential energy field* $P(T)$ is a scalar field defined on a flow poly-tree T, assigning each node v_i a potential p_{v_i}. For comparability, we normalize $P(T)$ so that its minimum value is zero by choosing an appropriate zero-potential reference.

Figure 3.2b and c visualize the construction of a basic potential field from a poly-tree, and Fig. 3.2d shows the normalized field after shifting the minimum to zero.

3.3.4 Inter-Community Potential Field (PEF) Modeling

The preceding methods primarily model potential fields within individual communities, which limits their applicability when flows span multiple communities. This section therefore extends the framework by constructing an inter-community PEF. To that end, we first define a *macroscopic flow graph* that describes the aggregated flow relations across communities.

Definition 3.9 (*Macroscopic Flow Graph*) The macroscopic flow graph $MG(V_{MG}, E_{MG})$ is a directed graph in which V_{MG} denotes the set of communities (one node per community), and E_{MG} denotes aggregated flows between adjacent communities, represented as directed edges.

As in the microscopic setting, the macroscopic flow graph may still exhibit *bidirectional flows* and *circulations* (see Fig. 3.2b). We therefore propose two complementary procedures to address these issues respectively: (i) a *major/minor decomposition* for removing bidirectionality, and (ii) a *general poly-tree decomposition* to eliminate cycles.

For any bidirectional community pair, the major/minor decomposition compares the magnitudes of the two directions and assigns the larger-direction edge to the *major* subgraph and the smaller-direction edge to the *minor* subgraph. This yields two subgraphs that share the same node set as MG but each retains only one direction of every bidirectional pair, effectively removing bidirectionality.

To further eliminate circulations, we introduce the General Polytree Decomposition (GPTD) algorithm, which operates on a directed macroscopic subgraph without bidirectional edges (i.e., either the major or the minor subgraph) and decomposes it into multiple edge-disjoint polytrees; see Algorithm 2. GPTD builds upon an adapted Kruskal minimum-spanning-tree procedure [8] to extract a tree from a directed acyclic subgraph by repeatedly:

(i) initializing each node as an individual tree,
(ii) selecting the minimum-weight edge that connects two distinct trees,

Algorithm 2: General Poly-Tree Decomposition (GPTD)

Input: Macroscopic flow graph $MG(V_{MG}, E_{MG})$ (with no bidirectional edges)
Output: A set MT of edge-disjoint flow poly-trees

Initialize visit counters $t_{uv} \leftarrow 0$ for all $e_{uv} \in E_{MG}$;
Initialize edge weights $w_{uv} \leftarrow 1/e_{uv}$ for all $e_{uv} \in E_{MG}$;
repeat
  Compute expansion factor $\phi \leftarrow w_{\max}/w_{\min} + 1$;
  Treat every node in MG as an individual tree;
  repeat
    Extract the edge e_{uv} with the smallest weight from E_{MG};
    if *u and v belong to different trees* **then**
      Merge the two trees and add e_{uv} to the merged tree;
  until *all nodes belong to a single tree T*;
  Add T to MT;
  for $e_{uv} \in E_T$ **do**
    $t_{uv} \leftarrow t_{uv} + 1$;
    $w_{uv} \leftarrow w_{uv} \times \phi \times 2^{t_{uv}}$;
until *$t_{uv} > 0$ holds for all $e_{uv} \in E_{MG}$*;
Distribute each edge's flow evenly across all poly-trees that contain it according to t_{uv};
return MT

(iii) merging the corresponding trees via that edge
(iv) iterating until all nodes belong to a single tree, yielding a cycle-free spanning structure.

Since Kruskal-style merging only adds edges between distinct trees, the resulting structure is guaranteed to be acyclic, thereby removing circulations. In GPTD, we set edge weights as the inverse of flow,

$$w_{uv} = \frac{1}{e_{uv}}, \quad \forall\, e_{uv} \in E_{MG}, \tag{3.13}$$

so that higher-flow connections are preferred and the resulting tree structure better reflects macroscopic flow patterns.

GPTD extracts multiple polytrees by repeatedly executing this Kruskal step until every input edge has appeared in at least one extracted tree (Algorithm 2, final lines). If edge weights were kept fixed, each run would return the same tree. To diversify the decomposition, GPTD updates edge weights dynamically after each run (Algorithm 2, inner loop). Let E_T be the edge set of the tree found at run i. For every $e_{uv} \in E_T$, we update its weight as

$$w_{uv}^{(i+1)} = w_{uv}^{(i)} \times \phi^{(i)} \times 2^{t_{uv}^{(i)}}, \tag{3.14}$$

where the expansion factor is

$$\phi^{(i)} = \frac{w_{\max}^{(i)}}{w_{\min}^{(i)}} + 1, \tag{3.15}$$

with $w_{\max}^{(i)}$ and $w_{\min}^{(i)}$ denoting the maximum and minimum edge weights at run i, respectively. This expansion pulls the weights of the just-selected edges toward the current maximum, thereby suppressing their probability of being re-selected in the next run. The multiplicative factor $2^{\cdot t_{uv}^{(i)}}$, where $t_{uv}^{(i)}$ counts how many times e_{uv} has been selected up to run i, further penalizes frequently selected edges, reducing overlaps and promoting diversity across the extracted polytrees.

The procedure repeats until every edge satisfies $t_{uv} > 0$, meaning that each edge has been included in at least one polytree. Finally, flows on shared edges are evenly distributed among the polytrees that contain them, yielding the final set MT. Based on these macroscopic polytrees, we then construct inter-community potential fields using the same PEF-generation principles as for microscopic trees, thereby completing the PEF extraction module across both intra- and inter-community scales.

3.4 Regional Traffic Flow Prediction Model

The goal of traffic flow prediction is to infer future flow states from decomposed historical PEF sequences. We adopt a two-stage paradigm: first, a spatiotemporal deep model predicts future PEFs; second, the predicted PEFs are mapped to future flow states.

In the first stage, the model takes historical PEF time series as input and predicts the next-step PEF. Because PEF sequences exhibit complex spatiotemporal dependencies, different neural components are designed to specialize in different dependency types. Concretely, we propose a correlation-adaptive spatiotemporal model composed of a temporal module and a spatial module. The temporal module jointly captures short-term and long-term temporal dependencies (see Fig. 3.4), enabling the data-driven backbone to exploit multi-scale correlations in PEF sequences for more accurate and robust prediction.

3.4.1 Temporal Modeling Component

Temporal dependencies are divided into two categories:

- *Short-term dependence*: autoregressive effects whereby the current state strongly depends on recent observations.
- *Long-term dependence*: periodic/seasonal patterns such as diurnal or weekly cycles.

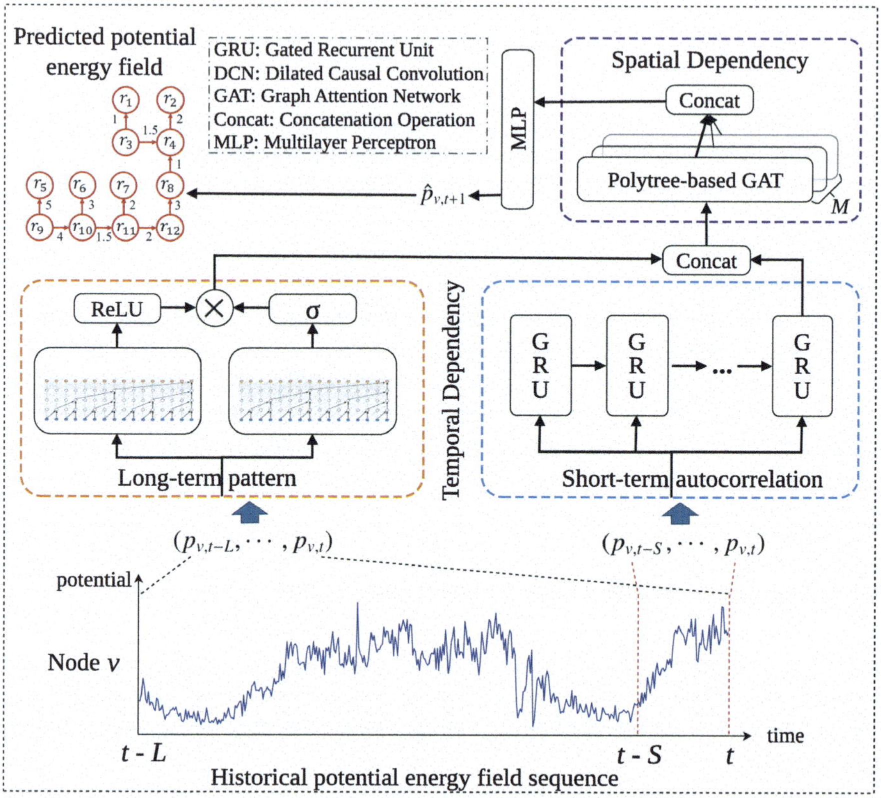

Fig. 3.4 Overview of the spatiotemporal architecture for PEF prediction

As observed in prior work [26], long-term periodic knowledge tends to generalize better and provides stable trend guidance, whereas short-term dependence is crucial at fine temporal resolutions for capturing recent fluctuations.

Accordingly, we employ distinct neural operators for the two temporal regimes: a recurrence-based model (e.g., a gated recurrent unit without attention) for short-term dynamics, and a dilated causal convolutional network (DCN) for long-term periodic patterns. This dual-path design enables the model to jointly learn local dynamics and global regularities, thereby improving predictive accuracy and generalization.

3.4.1.1 Short-Term Modeling

Let the historical potential sequence for node v be

$$\boldsymbol{p}_v^S = (p_{v,t-S}, \ldots, p_{v,t}). \tag{3.16}$$

We encode short-term dynamics using a gated recurrent unit (GRU) [4]:

$$\boldsymbol{h}_{v,t} = \mathrm{GRU}(\boldsymbol{h}_{v,t-1},\ p_{v,t}), \tag{3.17}$$

where $\boldsymbol{h}_{v,t} \in \mathbb{R}^{K_S}$ denotes the hidden temporal representation at time t. By explicitly modeling the recursion over recent steps, the GRU captures local autoregressive behavior that is key to short-term traffic variation.

3.4.1.2 Long-Term Modeling

Urban traffic typically exhibits strong long-horizon regularities (e.g., daily peaks) that may diverge substantially from the immediate past. RNNs often struggle with such long-range dependencies due to vanishing/exploding gradients. We therefore employ a dilated causal convolutional network (DCN) [17, 27] to model long-term patterns, as also advocated in [20, 25]. DCN uses dilated convolutions to expand the receptive field exponentially with depth while preserving causality (via zero padding), preventing information leakage from the future.

Given a long-horizon sequence $\boldsymbol{p}_v^L = (p_{v,t-L}, \ldots, p_{v,t}) \in \mathbb{R}^L$ and a 1-D filter $\boldsymbol{\theta} \in \mathbb{R}^r$, the dilated causal operator is

$$\mathcal{D}(\boldsymbol{\theta}, \boldsymbol{p}_v^L) = \sum_{i=0}^{r-1} \theta_i \cdot p_{v,\,t-d\cdot i}, \tag{3.18}$$

where d is the dilation factor. We adopt a hierarchical dilation schedule (increasing d layer by layer) to achieve an exponentially growing receptive field with comparatively few layers, and use multiple parallel filters to learn multi-scale periodicities.

In practice, we feed the potential sequences from the past three days into the DCN to capture long-term dependencies. Since traffic regimes differ markedly across time (e.g., peak vs. off-peak, weekday vs. weekend), we further introduce a *gating mechanism* that adaptively activates the appropriate regime while suppressing irrelevant ones:

$$\boldsymbol{c}_{v,t} = \mathrm{ReLU}\big(\mathrm{DCN}(\boldsymbol{\Theta}_1, \boldsymbol{p}_v^L)\big) \odot \sigma\big(\mathrm{DCN}(\boldsymbol{\Theta}_2, \boldsymbol{p}_v^L)\big), \tag{3.19}$$

where $\boldsymbol{c}_{v,t} \in \mathbb{R}^{K_L}$ is the long-term representation at time t, σ is the sigmoid function, $\odot$ denotes element-wise multiplication, and $\boldsymbol{\Theta}_1, \boldsymbol{\Theta}_2 \in \mathbb{R}^{N_L \times K_L \times r}$ are learnable filter banks over N_L DCN layers. The second term serves as a gate on the first term's information flow.

Compared with recurrence-based approaches, DCN offers several advantages:

- *Alleviates gradient issues*: Dilated convolutions directly encode long-distance dependencies without long recurrent chains.
- *Efficient long-range modeling*: Exponentially expanding receptive fields capture long-horizon patterns with modest depth.

- *Gated adaptivity*: The gating path dynamically selects regimes (e.g., peak vs. off-peak), improving robustness to heterogeneous temporal patterns.

Relative to multi-scale handcrafted history construction [31, 32], DCN reduces manual effort and automates long-te

3.4.2 Spatial Modeling Component

To capture the dynamic spatial dependencies embedded in PEFs, we extend the Graph Attention Network (GAT) [21] and propose a *Polytree-Based GAT* (PB-GAT). We detail its computation below.

We begin with the standard GAT update. Let $\boldsymbol{n}_{v_k}$ denote the representation of PEF node v_k. At time t, GAT iteratively updates a matrix $\boldsymbol{N} \in \mathbb{R}^{K_G \times |V_T|}$ that stacks the representations of all PEF nodes:

$$\boldsymbol{N}^{(z)} = \text{GAT}\big(\boldsymbol{N}^{(z-1)} \mid t\big), \tag{3.20}$$

where z indexes the iteration. To inject temporal information into the spatial module, we initialize node representations as $\boldsymbol{n}_v^{(0)} = [\boldsymbol{h}_{v,t}, \boldsymbol{c}_{v,t}]$, where $\boldsymbol{h}_{v,t}$ and $\boldsymbol{c}_{v,t}$ are the short-term autoregressive and long-term periodic components for node v at time t, respectively.

Concretely, the original GAT defines the attention weight between nodes v_i and v_j as

$$\alpha_{(v_i,v_j)} = \frac{\exp\big(\boldsymbol{w}^\top(\boldsymbol{W}_1\boldsymbol{n}_{v_i} + \boldsymbol{W}_2\boldsymbol{n}_{v_j})\big)}{\sum_{v_k \in \mathcal{N}_{v_i}} \exp\big(\boldsymbol{w}^\top(\boldsymbol{W}_1\boldsymbol{n}_{v_i} + \boldsymbol{W}_2\boldsymbol{n}_{v_k})\big)}, \tag{3.21}$$

where $\alpha_{(\cdot)}$ is the attention score, $\boldsymbol{w}$ and $\boldsymbol{W}_{(\cdot)}$ are learnable parameters, and $\mathcal{N}_{v_i}$ denotes the neighbor set of v_i.

The original attention formulation in Eq. (3.21) implicitly assumes an undirected input graph. In our setting, however, the PEFs defined on the corresponding poly-trees are directed, and spatial influences between two nodes (representing different locations) are typically asymmetric. To accommodate this asymmetry, we introduce a *weighted, directed attention* mechanism into the graph attention network (GAT). Specifically, for each node v_i, we define two attention scores:

$$\begin{aligned}
\alpha_{(v_i,v_j)} &= \frac{\exp\big(e_{ij}\,\boldsymbol{w}^\top(\boldsymbol{W}_1\boldsymbol{n}_{v_i} + \boldsymbol{W}_2\boldsymbol{n}_{v_j})\big)}{\sum_{v_k \in \mathcal{LO}_{v_i}} \exp\big(e_{ik}\,\boldsymbol{w}^\top(\boldsymbol{W}_1\boldsymbol{n}_{v_i} + \boldsymbol{W}_2\boldsymbol{n}_{v_k})\big)},\\
\alpha_{(v_j,v_i)} &= \frac{\exp\big(e_{ji}\,\boldsymbol{w}^\top(\boldsymbol{W}_1\boldsymbol{n}_{v_i} + \boldsymbol{W}_2\boldsymbol{n}_{v_j})\big)}{\sum_{v_k \in \mathcal{HI}_{v_i}} \exp\big(e_{ki}\,\boldsymbol{w}^\top(\boldsymbol{W}_1\boldsymbol{n}_{v_i} + \boldsymbol{W}_2\boldsymbol{n}_{v_k})\big)},
\end{aligned} \tag{3.22}$$

where $e_{(\cdot)}$ is the edge weight, and $\mathcal{LO}_{v_i}$ and $\mathcal{HI}_{v_i}$ denote, respectively, the neighbors of v_i with lower and higher potentials on the poly-tree structure.

We then apply multi-head attention to update each node representation using the directed scores in Eq. (3.22), thereby enhancing the model's ability to capture spatial heterogeneity:

$$\boldsymbol{n}_{v_i}^{(z)} = \Big\Vert_{m=1}^{M} \mathrm{ReLU}\left(\sum_{v_k \in \mathcal{LO}_{v_i}} \alpha_{(v_i,v_k)}^{(m)} \boldsymbol{W}^{(m)} \boldsymbol{n}_{v_k}^{(z-1)} + \sum_{v_k \in \mathcal{HI}_{v_i}} \alpha_{(v_k,v_i)}^{(m)} \boldsymbol{W}^{(m)} \boldsymbol{n}_{v_k}^{(z-1)} \right), \tag{3.23}$$

where $\Vert$ denotes concatenation, ReLU is the rectified linear unit, M is the number of attention heads, $\alpha_{(\cdot)}^{(m)}$ is the normalized attention from head m, and $\boldsymbol{W}^{(m)}$ is learnable. Because the extended GAT explicitly accounts for edge directions and weights on the underlying poly-trees during attention computation, we refer to it as *PB-GAT*.

3.4.3 Model Training and Flow Inference

Since the deep model predicts *future PEF states*, we must convert predicted potentials into future road-network flows. We first describe the training objective for the spatio-temporal model, and then explain how to infer traffic flows from predicted PEFs.

Given the learned representation $\boldsymbol{n}_{v,t}$, the potential at time $t+1$ for poly-tree node v is predicted as

$$\hat{p}_{v,t+1} = \mathrm{MLP}(\boldsymbol{n}_{v,t}), \tag{3.24}$$

where MLP denotes a multilayer perceptron. We train the model by minimizing the mean squared error (MSE):

$$\mathcal{L} = \sum_{v \in V_T} \left(p_{v,t+1} - \hat{p}_{v,t+1}\right)^2. \tag{3.25}$$

Repeating this process over all communities yields community-wise PEF predictions.

3.4.3.1 Training Strategy

In the PEF extraction module, the flow graph is decomposed into multiple polytrees. A straightforward but computationally expensive strategy would be to train a separate model for each polytree. To improve efficiency, we adopt a *hybrid* parameter-sharing strategy: in the temporal component, nodes across different communities share parameters, reflecting broadly similar temporal patterns across locations; in the spatial component, polytrees with the same backbone direction across communities share parameters, exploiting structural commonalities across PEFs. This design largely preserves predictive accuracy while substantially reducing training and inference costs.

Algorithm 3: Flow-Graph Generation from Potential Fields

```
Input: List of potential fields P, community set C
Output: City-level flow graph G
Initialize G to zero on all nodes and edges;
for c ∈ C do
    Initialize the community flow graph G_c to zero;
    for p ∈ P do
        Initialize a temporary flow graph G_t;
        for (u, v) ∈ p do
            if p_u > p_v then
                G_t(u, v) ← p_u − p_v;
            else
                G_t(v, u) ← p_v − p_u;
        Accumulate G_t into G_c;
    Retrieve the boundary of c;
    Accumulate G_c (including boundary edges) into G;
return G
```

3.4.3.2 Flow Inference Component

After predicting the PEFs, we reconstruct traffic flows at the community level and then aggregate them into a citywide flow map; see Algorithm 3. For each PEF edge, the predicted flow is computed from the potential difference between its endpoints. Iterating over all edges in a PEF yields the corresponding flow polytree. Finally, we merge community-level flow polytrees into a unified urban flow graph. Because adjacent communities share boundary nodes, we average the flows on boundary edges across all relevant communities to obtain coherent and consistent boundary estimates.

3.5 Empirical Evaluation and Findings

This section presents a comprehensive empirical evaluation of the proposed methods. It introduces the evaluation setup, followed by a comparative analysis against representative baselines. The section further examines the contribution of key model components through ablation studies and provides an in-depth analysis of model behavior to assess interpretability, robustness, and generalization.

3.5.1 Evaluation Setup

This subsection describes the experimental settings used for empirical evaluation, including the characteristics of the datasets, the evaluation metrics and baseline methods for comparison, and key implementation details.

3.5.1.1 Data Description

We evaluate STPEF on three real-world, grid-based traffic forecasting datasets: Xi'an (XA), Beijing (BJ), and Porto (PO). Each dataset consists of raw taxi GPS trajectories with different sampling intervals. The study area in each city is partitioned into grid cells, and inter-region flows are computed from trajectories as follows:

- *Xi'an taxi dataset* (from the DiDi GAIA open data) contains 3,784,063 trajectories sampled every 3 seconds, spanning 2016-10-01 to 2016-10-31. We partition Xi'an into 550 m $\times$ 500 m grids and aggregate inter-region flows. The first 25 days are used for training and the remainder for testing.
- *Beijing taxi dataset* contains 60,828,387 trajectories with a 1-minute sampling interval, covering two periods: 2015-01-01 to 2015-02-28 and 2015-04-01 to 2015-07-31. The Beijing area is partitioned into 800 m $\times$ 600 m grids. We use the most recent five weeks for testing and the rest for training.
- *Porto taxi dataset* was released by a Kaggle trajectory prediction competition (15-second sampling).[3] It covers 2013-07-01 to 2014-06-30 with 1,891,921 trajectories. Porto is partitioned into 500 m $\times$ 500 m grids; data from 2013-07-01 to 2014-04-15 are used for training and the remainder for testing.

To reflect heterogeneous urban dynamics, we adopt different temporal aggregation intervals across datasets: 5 min for Xi'an, 30 minutes for Beijing, and 60 minutes for Porto. Detailed statistics are summarized in Table 3.1.

The grid resolutions also differ across Xi'an, Beijing, and Porto owing to differences in city size, urban structure, and spatial flow distribution. As a megacity, Beijing exhibits broader and more complex inter-regional flows; therefore, we adopt larger grid cells to avoid excessive noise caused by overly fine partitions. Xi'an and Porto, as medium-sized cities, have relatively concentrated flows, so smaller grids are used to better capture local patterns. Although the grid resolution affects the number of nodes in the flow graph, we subsequently perform community detection on the grid graph and control the target number of communities, keeping it relatively stable across different resolutions.

[3] Dataset URL: https://www.kaggle.com/c/pkdd-15-predict-taxi-service-trajectory-i.

Table 3.1 Summary statistics of the Xi'an, Beijing, and Porto taxi datasets

City	Xi'an (XA)	Beijing (BJ)	Porto (PO)
# Trajectories	3,784,063	60,828,387	1,891,921
GPS sampling interval	3 s	60 s	15 s
Coverage	31 days	181 days	365 days
Grid size	550 m × 500 m	800 m × 600 m	500 m × 500 m
Flow interval	5 min	30 min	60 min

3.5.1.2 Evaluation Metrics and Baselines

We report four common error-based metrics: mean absolute error (MAE), root mean squared error (RMSE), coefficient of determination (R^2), and explained variance score (EVAR). Larger R^2 and EVAR indicate better goodness-of-fit and variance explanation, respectively.

We compare STPEF against nine baselines spanning five categories.

- Traditional time-series models:
  - HA: Historical Average with seasonality. We average the same time-of-week from the previous four weeks as the prediction.
  - ARIMA: Autoregressive Integrated Moving Average with order selected on the validation set; we use (3, 0, 1).
- RNN-based deep forecasting:
  - FC-LSTM [18]: An encoder-decoder model with fully connected LSTM units. Both encoder and decoder contain one FC layer with 256 hidden units.
- CNN-based spatio-temporal forecasting:
  - ST-ResNet [32]: Uses multiple temporal sequences to capture temporal dependencies and a residual network to model spatial correlations.
  - STDN [29]: Integrates LSTM, local CNN, and semantic embeddings to jointly learn spatio-temporal dependencies.
- GNN-based traffic forecasting:
  - STGCN [30]: A purely convolutional approach combining graph convolution (space) with 1D CNN (time).
  - AGCRN [1]: Augments GCN with an adaptive module for learnable adjacency and embeds it into an RNN to capture spatio-temporal dependencies.
- Attention-based approaches:
  - STAWnet [20]: Employs self-attention to learn node embeddings for spatial dependencies and dilated convolutions for temporal modeling.

– STFGNN [9]: Fuses a data-generated temporal graph with a predefined spatial graph to extract hidden spatio-temporal dependencies.

We additionally construct an *ablated* baseline that differs from our full model in two key aspects: (i) the temporal component ignores long-term periodicity and only models short-term autoregression via GRU; and (ii) the spatial component ignores the directed polytree structure and instead employs a standard GAT. This baseline is designed to isolate and highlight the contributions of periodicity and directionality. For clarity, we denote our full model as **STPEF+** and the ablated variant as **STPEF**.

All deep-learning baselines are implemented and executed within the LibCity platform [23], a unified and extensible library that provides standardized implementations and ensures fair and reproducible comparisons. For each method, we follow the hyperparameter settings from the corresponding official code releases whenever available, and tune the remaining hyperparameters on the validation set.

3.5.1.3 Implementation Details

In **STPEF+**, traffic *flows* are defined on the *edges* of the grid graph, whereas most baselines are formulated for *node*-level prediction (typically using two regression heads for inflow and outflow). To adapt node-based baselines to edge-level prediction, we extend their output layer to four regression heads corresponding to outflows toward the up, down, left, and right neighbors. Since each node's outflow equals the inflow of its neighbor, node-based baselines remain compatible with edge-level evaluation, enabling a fair comparison with **STPEF+**.

We focus on micro-level predictions, as they directly align with Definition 3.4. Macroscopic PEF predictions primarily serve to interpret inter-community distributions and long-range commuting patterns. We consider three forecasting horizons—predicting the next 1, 2, and 4 future steps—to jointly assess short-term and medium-/long-term performance. For multi-step prediction, we adopt a recursive strategy [19], feeding the prediction at time t as part of the input for time $t+1$ iteratively.

We split the training data into 90% for parameter optimization and 10% for validation and hyperparameter tuning. All learnable parameters are initialized from a truncated normal distribution (mean 0, variance 0.01), and all biases are zero-initialized. We use Adam [7] to minimize MSE, with early stopping to mitigate overfitting. The learning rate is set to 0.001, and both the maximum number of epochs and the mini-batch size are set to 100.

For Xi'an, Beijing, and Porto, the numbers of communities are 8, 12, and 4, respectively. The number of components in the flow polytree decomposition is fixed to 4. The GRU input length is 4. The DCN input lengths are 864, 144, and 72; the numbers of DCN layers are 8, 6, and 5; and the dilation at DCN layer n_L is 2^{n_L-1}. The DCN filter size is selected from {4, 5, 6, 7}. The number of GAT layers is selected from {1, 2, 3, 4, 5}, the number of attention heads from {1, 2, 4, 8, 16, 32}, and the hidden dimensionality from {64, 128, 256, 512}.

All experiments are conducted on a server equipped with 64 GB RAM, dual Intel Xeon E5-2680 v2 CPUs, and one NVIDIA GeForce GTX 1080 Ti GPU, running Ubuntu 16.04. We use Python 3.6.10 and PyTorch 1.4.0 with CUDA 9.2.

3.5.2 Overall Comparative Analysis

Tables 3.2, 3.3, and 3.4 report the results of different forecasting methods on the Xi'an, Beijing, and Porto taxi datasets, respectively. Because the absolute metric values are small, we keep three digits after the decimal point for ease of comparison. To ensure *stability* and *reproducibility*, each experimental configuration is repeated 15 times and we report the mean performance across runs. The results highlight three main findings.

Table 3.2 Regional traffic flow forecasting on the Xi'an dataset

Metric	MAE ↓			RMSE ↓			R^2 ↑			EVAR ↑		
Horizon	5 min	10 min	20 min	5 min	10 min	20 min	5 min	10 min	20 min	5 min	10 min	20 min
HA	2.032	2.032	2.032	3.840	3.840	3.840	0.618	0.618	0.618	0.616	0.616	0.616
ARIMA	1.987	2.253	2.586	3.763	4.258	4.861	0.621	0.607	0.584	0.620	0.606	0.584
FC-LSTM	1.602	1.975	2.324	2.884	3.653	4.323	0.706	0.625	0.595	0.708	0.625	0.596
ST-ResNet	1.102	1.208	1.322	1.725	1.943	2.208	0.810	0.795	0.782	0.811	0.797	0.780
STDN	1.252	1.326	1.439	2.018	2.147	2.325	0.797	0.790	0.778	0.798	0.790	0.779
STGCN	1.141	1.255	1.422	1.898	2.025	2.314	0.809	0.796	0.781	0.810	0.796	0.782
AGCRN	1.105	1.208	1.373	1.854	1.969	2.271	0.811	0.801	0.785	0.811	0.802	0.785
STAWnet	1.008	1.120	1.269	1.766	1.933	2.181	0.816	0.810	0.793	0.816	0.809	0.794
STFGNN	0.987	1.093	1.233	1.708	1.847	2.009	0.821	0.813	0.800	0.822	0.811	0.798
STPEF	1.012	1.110	1.207	1.783	1.867	1.968	0.818	0.809	0.798	0.819	0.810	0.798
STPEF+	**0.859**	**0.921**	**1.003**	**1.595**	**1.663**	**1.759**	**0.840**	**0.829**	**0.818**	**0.842**	**0.829**	**0.819**

Table 3.3 Regional traffic flow forecasting on the Beijing dataset

Metric	MAE ↓			RMSE ↓			R^2 ↑			EVAR ↑		
Horizon	30 min	1 h	2 h	30 min	1 h	2 h	30 min	1 h	2 h	30 min	1 h	2 h
HA	5.241	5.241	5.241	10.587	10.587	10.587	0.635	0.635	0.635	0.634	0.634	0.634
ARIMA	5.125	5.831	6.631	10.301	11.735	13.160	0.643	0.616	0.587	0.642	0.615	0.588
FC-LSTM	4.548	5.112	6.043	9.141	10.478	12.283	0.670	0.641	0.603	0.671	0.640	0.601
ST-ResNet	2.583	3.162	3.975	5.159	6.784	8.826	0.853	0.814	0.751	0.854	0.816	0.753
STDN	3.415	4.342	5.512	7.210	9.392	11.578	0.805	0.687	0.622	0.804	0.687	0.620
STGCN	3.374	4.340	5.418	7.122	9.380	11.322	0.809	0.686	0.627	0.809	0.687	0.625
AGCRN	3.018	3.839	5.152	6.763	8.512	10.793	0.821	0.745	0.640	0.820	0.746	0.640
STAWnet	2.602	3.034	3.866	5.316	6.691	8.545	0.860	0.823	0.742	0.861	0.823	0.741
STFGNN	2.563	2.730	2.945	5.169	5.793	6.566	0.867	0.851	0.832	0.867	0.852	0.833
STPEF	2.572	2.731	2.986	5.173	5.762	6.647	0.865	0.855	0.834	0.864	0.854	0.836
STPEF+	**2.186**	**2.375**	**2.518**	**4.590**	**4.941**	**5.326**	**0.904**	**0.883**	**0.870**	**0.905**	**0.885**	**0.871**

Table 3.4 Regional traffic flow forecasting on the Porto dataset

Metric	MAE ↓			RMSE ↓			R^2 ↑			EVAR ↑		
Horizon	1 h	2 h	4 h	1 h	2 h	4 h	1 h	2 h	4 h	1 h	2 h	4 h
HA	2.610	2.610	2.610	4.672	4.672	4.672	0.532	0.532	0.532	0.531	0.531	0.531
ARIMA	2.573	2.862	3.288	4.606	5.209	6.017	0.539	0.514	0.440	0.539	0.514	0.442
FC-LSTM	2.014	2.469	3.131	3.473	4.572	5.793	0.626	0.551	0.479	0.624	0.552	0.480
ST-ResNet	1.594	1.705	1.853	2.664	2.842	3.168	0.745	0.711	0.670	0.745	0.711	0.668
STDN	1.748	1.946	2.181	2.903	3.242	3.806	0.701	0.653	0.598	0.702	0.654	0.597
STGCN	1.785	1.950	2.179	2.936	3.251	3.803	0.695	0.650	0.598	0.695	0.651	0.596
AGCRN	1.720	1.856	2.087	2.886	3.127	3.642	0.700	0.674	0.632	0.699	0.673	0.630
STAWnet	1.625	1.701	1.814	2.793	2.870	3.061	0.730	0.705	0.680	0.731	0.704	0.680
STFGNN	1.604	1.679	1.761	2.726	2.834	2.975	0.738	0.718	0.689	0.739	0.717	0.688
STPEF	1.612	1.751	1.826	2.786	2.910	3.106	0.731	0.697	0.676	0.730	0.696	0.675
STPEF+	**1.408**	**1.513**	**1.610**	**2.449**	**2.592**	**2.758**	**0.802**	**0.770**	**0.735**	**0.800**	**0.769**	**0.735**

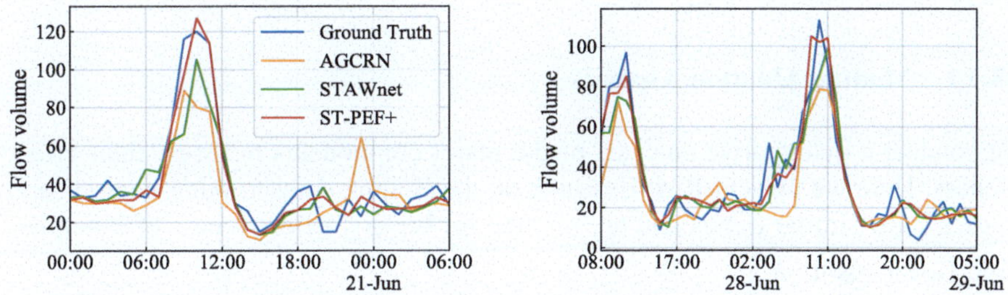

Fig. 3.5 Peak-hour forecasting cases on the Porto taxi dataset

First, paired t-tests indicate that **STPEF+** significantly outperforms all baselines at the 0.01 significance level, highlighting the effectiveness of coupling potential-field decomposition with correlation-adaptive deep models. Figure 3.5 visualizes two randomly sampled forecasting cases to illustrate where these gains arise, where *Ground Truth* denotes the realized future traffic flow. As shown, **STPEF+** tracks peak-hour trends more accurately than the competing methods. Because traffic dynamics exhibit a larger amplitude during peak periods, accurately forecasting these segments is crucial for overall performance.

In **STPEF+**, directional flows are decomposed into independent PEFs and modeled by distinct networks. During peak hours, when traffic is often dominated by a single direction, this design enables the model to capture changes more precisely. This mechanism likely explains the consistent advantage of **STPEF+** over the baselines.

Second, relative to **STPEF**, **STPEF+** achieves an average improvement of 12.4% in RMSE.This suggests that the advanced modules—PB-GAT with directional attention and the gated DCN for long-term patterns—play a key role in the superior performance of **STPEF+**. Without these enhanced spatiotemporal components (i.e. using **STPEF** alone), the model struggles to outperform strong recent baselines such as STFGNN and STAWnet.

Third, methods that explicitly model *spatiotemporal* dependencies generally outperform traditional time-series approaches and vanilla RNNs, underscoring the importance of spatiotemporal structure for traffic forecasting. Among these, GNN-based models are overall the most competitive, validating the incorporation of graph structure. Moreover, attention mechanisms further improve predictive accuracy by adaptively weighting heterogeneous spatial contexts.

3.5.3 Ablation Studies on Key Modules

This section mainly analyzes the rationality of the model design and the sensitivity of hyperparameters.

3.5.3.1 Module Ablation Analysis

We conduct a quantitative analysis of the major components to assess their effectiveness. Here we present RMSE results on the Beijing dataset; other datasets and metrics show similar trends and are omitted for brevity. Effectiveness of Potential-

Field Decomposition
We first evaluate the PEF decomposition, which is the core component shared by STPEF+ and STPEF. By removing bidirectional and cyclic patterns from the traffic-flow graph, PEF makes pattern learning more direct. To validate this, we compare STPEF+ and STPEF with two tailored baselines, *ST-Flow* and *ST-Flow+*, which apply the same spatiotemporal modules as STPEF and STPEF+ directly on the *raw* flow graphs without PEF conversion. As shown in Fig. 3.6a, both *ST-Flow* and *ST-Flow+* underperform STPEF, indicating that PEF decomposition is critical to the observed gains.
Effectiveness of Temporal Modeling
We compare three variants of the temporal module: (i) *DCN*, which uses only a dilated causal convolutional network to capture long-term patterns (ignoring short-term autocorrelation); (ii) *GRU*, which only models short-term dependencies; and (iii) *Both*, which combines DCN and GRU.

The results in Fig. 3.6b show a consistent performance ranking *DCN* < *GRU* < *Both*. *GRU* outperforms *DCN*, indicating that short-term autocorrelation is often more critical than long-period patterns for traffic forecasting. The combined *Both* variant achieves the best performance, confirming the benefit of jointly modeling both short- and long-term temporal dependencies.

Effectiveness of Spatial Modeling
Fixing the temporal module to its best configuration, we further compare three spatial variants: (i) *N-GAT*, a variant without attention (node transformations are retained, but neighbor messages are aggregated with identical weights); (ii) *O-GAT*, the original GAT for undirected graphs; and (iii) *PB-GAT*, our enhanced GAT that encodes polytree context and edge information.

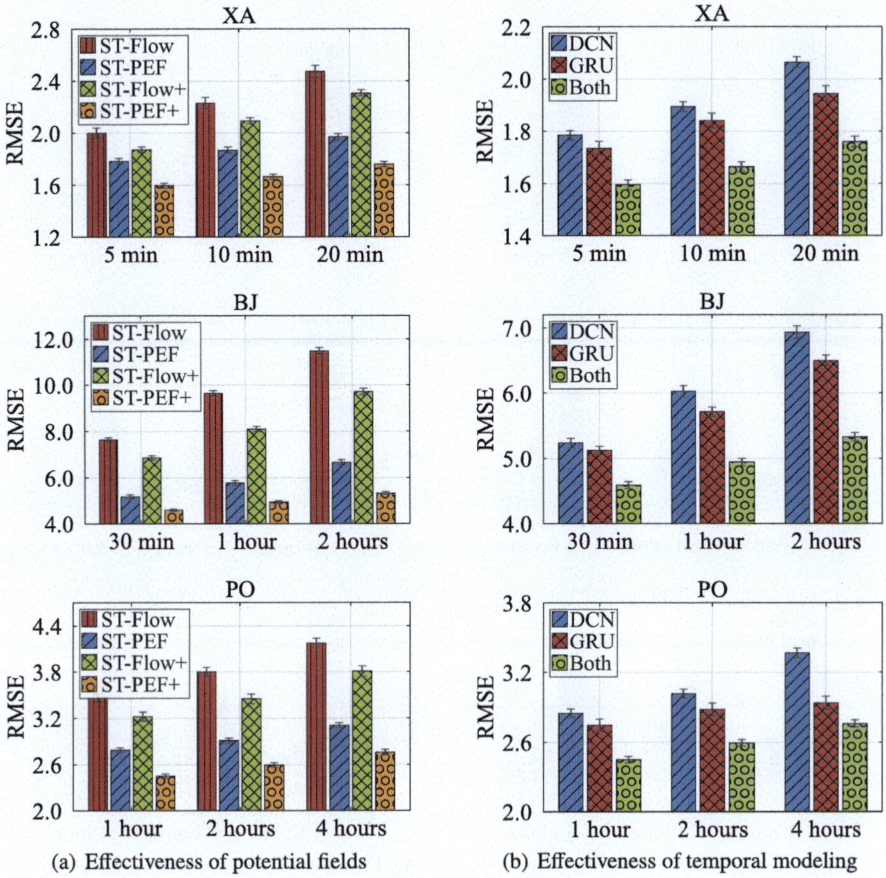

Fig. 3.6 Ablation on PEF and temporal modeling for STPEF+

As shown in Fig. 3.7a, the performance ranking is *N-GAT* < *O-GAT* < *PB-GAT*. The gap between *N-GAT* and the attention-based variants highlights the necessity of non-uniform neighbor weighting. *PB-GAT* consistently outperforms *O-GAT*, validating the utility of incorporating directed polytree edges into spatial dependency modeling.

In terms of efficiency, the two attention-based variants differ only in whether polytree directionality is used in the attention computation; the number of neighbors per node remains unchanged. Empirically, one training epoch takes 104.18 s for *O-GAT* and 104.48 s for *PB-GAT*, i.e., virtually identical. Thus, *PB-GAT* does not introduce noticeable computational overhead.

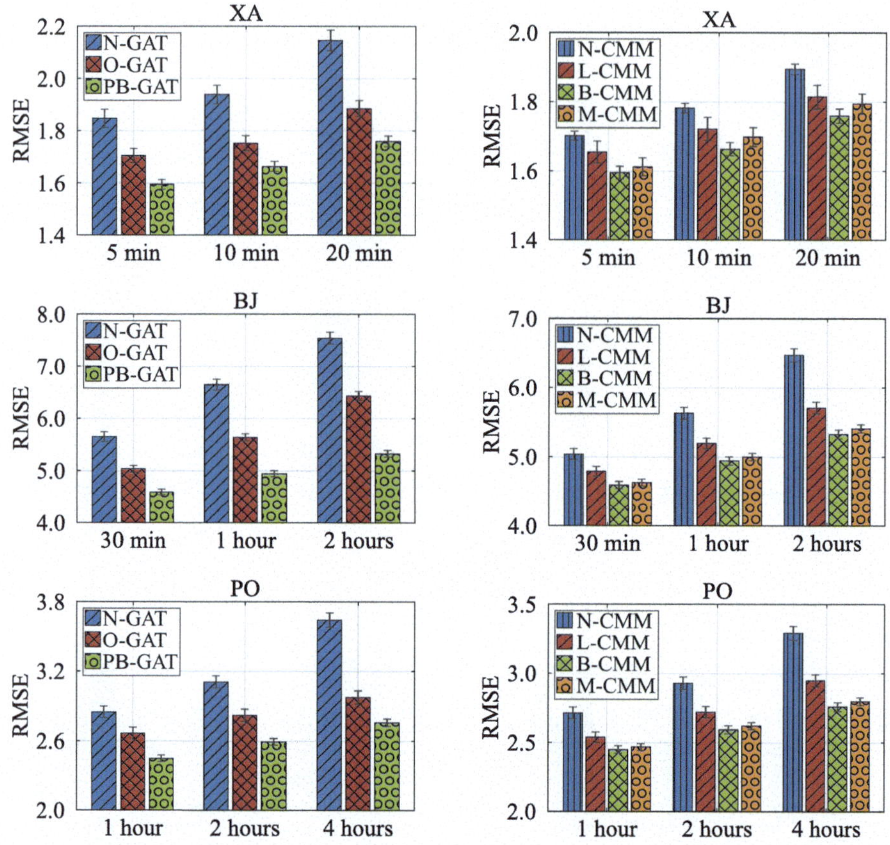

(a) Effectiveness of spatial modeling (b) Effectiveness of community partitioning

Fig. 3.7 Ablation on spatial modeling and community partitioning for STPEF+

Effectiveness of Community Partitioning

We assess four variants to evaluate the community module: (i) *N-CMM*, no partition (a single city-wide community); (ii) *L-CMM*, fewer-than-optimal communities, here $C = 6$ (optimal is $C = 12$); (iii) *B-CMM*, the best setting with $C = 12$; and (iv) *M-CMM*, more-than-optimal communities with $C = 15$.

Figure 3.7b shows the ranking *N-CMM* < *L-CMM* < *M-CMM* < *B-CMM*. *N-CMM* performs worst because city-wide dynamics are too complex to model directly, supporting a divide-and-conquer strategy. Among partitioned variants, *B-CMM* ($C = 12$) performs best. *L-CMM* likely merges distinct functional areas, entangling disparate patterns; *M-CMM* over-fragments communities, weakening spatial correlations and hurting performance.

3.5.3.2 Hyperparameter Sensitivity

We study the effects of DCN filter size, PB-GAT depth, PB-GAT head count, and the shared hidden dimensionality of GRU/DCN/PB-GAT. For simplicity, we tie the hidden size across the three modules and refer to it as the model hidden size. Results are shown in Fig. 3.8.

From Fig. 3.8, we conclude: (i) Varying the DCN filter size r shows that $r = 5$ yields the best results. (ii) Increasing GAT depth expands the receptive field and improves performance up to three layers; higher depths bring diminishing returns. (iii) Using eight attention heads best balances contextual richness and trainability; 16 or 32 heads degrade performance, likely due to increased parameterization and training difficulty. (iv) While larger hidden sizes enhance expressiveness, overly large sizes tend to overfit. We select 128 as a balanced choice between complexity and generalization.

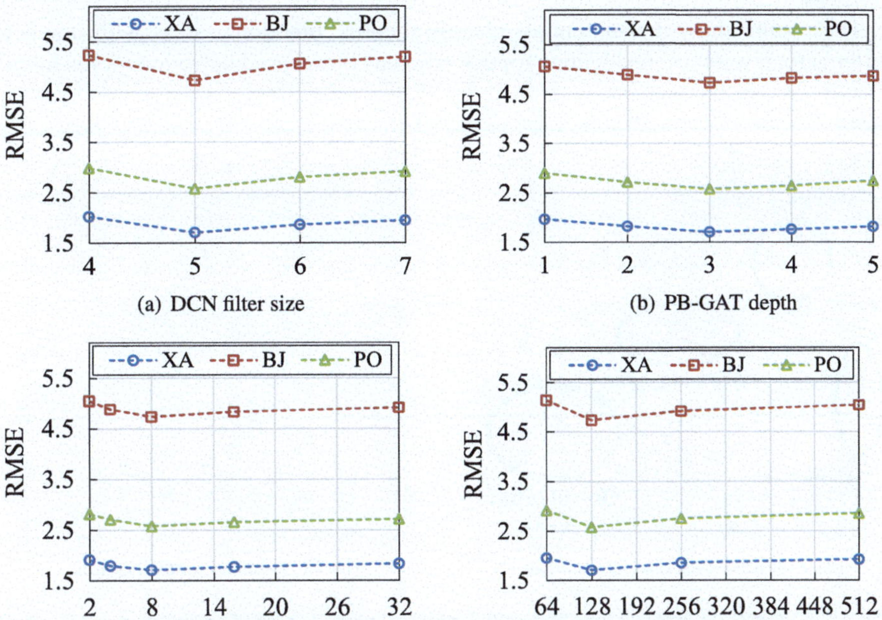

(a) DCN filter size (b) PB-GAT depth (c) PB-GAT head count (d) Hidden size

Fig. 3.8 Hyperparameter study for STPEF+

3.5.4 In-Depth Analysis of Model Behavior

We further visualize the learned potential fields on the Beijing dataset to analyze model behavior. The case study highlights two advantages: (1) PEFs effectively identify functional areas with distinct traffic patterns; and (2) PEFs capture the physical evolution of flows, offering interpretability.

3.5.4.1 Identifying Functional Areas via Potential Fields

We demonstrate the ability of learned PEFs to identify functional zones. Figure 3.9 presents spatial patterns mined by our approach during a weekday evening peak (17:30–19:30). In Fig. 3.9a, we compute the total potential of each node across all PEFs and mark the top 4% nodes in pink and the bottom 4% in yellow. For comparison, we compute the net outflow of each node and mark the top/bottom 4% in Fig. 3.9b.

The results show that potentials cleanly identify functional areas with distinct patterns (e.g. work vs. residential zones), whereas flow-based detection yields more

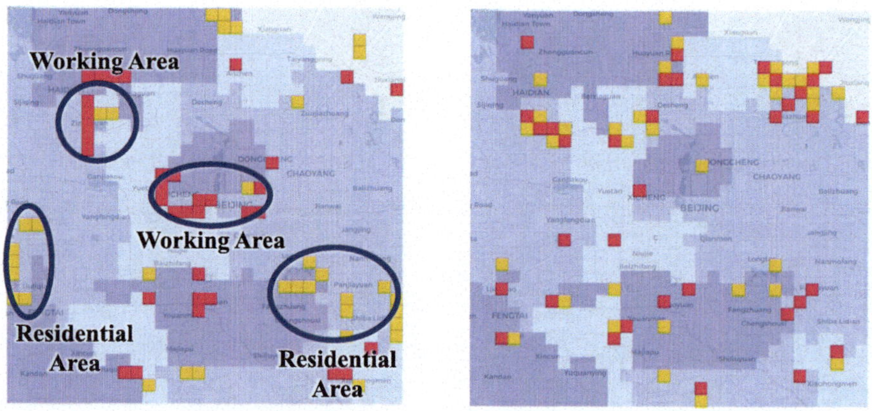

(a) Potential-based identification (b) Flow-based identification

Fig. 3.9 Comparison between potential- and flow-based identification of key areas scattered patterns. This is because the proposed notion of potential naturally characterizes *sources* and *sinks* in a dynamical system. In urban dynamics, driven by human mobility, work areas typically act as sinks in the morning peak and sources in the evening peak, while residential areas exhibit the opposite pattern [10, 28]. By decomposing flows, PEFs extract purer components, making pattern discovery clearer. In contrast, such patterns are often obscured in raw flow data due to the mixing of heterogeneous flow components, which complicates identification.

3.5.4.2 Interpreting Traffic-Flow Evolution via Potential Fields

This part demonstrates how the proposed PEF explains the physical evolution of traffic flows. In our framework, the city is partitioned into multiple communities, and PEFs are constructed at two levels: within communities and across communities. We therefore analyze the model's explanatory capacity at both levels.

We begin with within-community flows. In Fig. 3.10, we compute the total potential of each node on weekdays and weekends, highlighting high-potential nodes in pink and low-potential nodes in yellow. Figure 3.10a and b visualize the "tidal" potential effect revealed by the PEFs in urban Beijing.

During the weekday morning peak (Fig. 3.10a), potential-driven mobility typically proceeds from residential areas (pink, high potential) toward central business districts (yellow, low potential). In the evening peak (Fig. 3.10b), this pattern reverses as people leave workplaces and return to residential zones. Circles mark regions where potentials "flip" between morning and evening peaks, and arrows indicate the potential-driven direction of flow (from higher to lower potential).

On weekends, the citywide tidal effect is weaker, yet still discernible in specific regions such as commercial districts and transport hubs. For instance, Fig. 3.10c shows that commercial streets (e.g. Xidan, Wangfujing) and major railway stations (e.g. Beijing West Railway Station) exhibit low potentials in the morning, drawing inflows; by evening (Fig. 3.10d), these areas exhibit higher potentials, driving outflows. Some flows move back toward residential areas, while others head to entertainment districts (e.g. Sanlitun).

These observations indicate that PEFs effectively capture the spatiotemporal evolution of traffic, revealing how different functional areas shift in potential over time.

Figure 3.11 further plots potential trajectories for representative regions on weekdays and weekends. In Fig. 3.11a, during weekday mornings, the potential of a working district (region A in Fig. 3.10a) is lower than that of a residential district (region B), whereas the pattern reverses in the late afternoon, reflecting the change in commuting direction. In Fig. 3.11b, weekend potentials for commercial streets (e.g., Wangfujing, Xidan) and transport hubs (marked in Fig. 3.10c) exhibit regular cycles: potentials are low in the morning and then fluctuate three times between 10:00 and 22:00. By contrast, residential areas remain relatively stable at lower potentials, indicating that people predominantly stay at home on weekends and move in more regular rhythms.

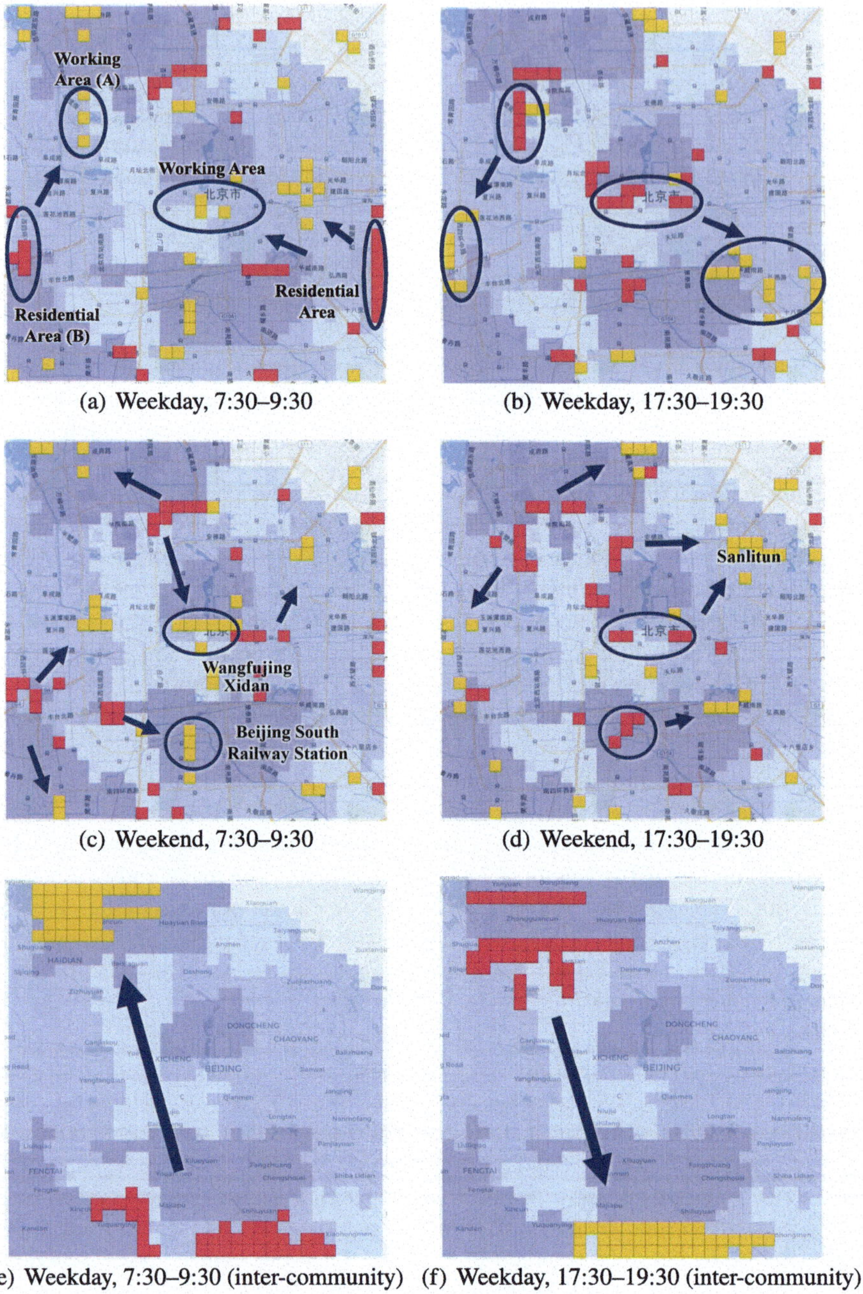

Fig. 3.10 Spatial distributions of PEFs within communities (subfigures (a)–(d)) and across communities (subfigures (e)–(f))

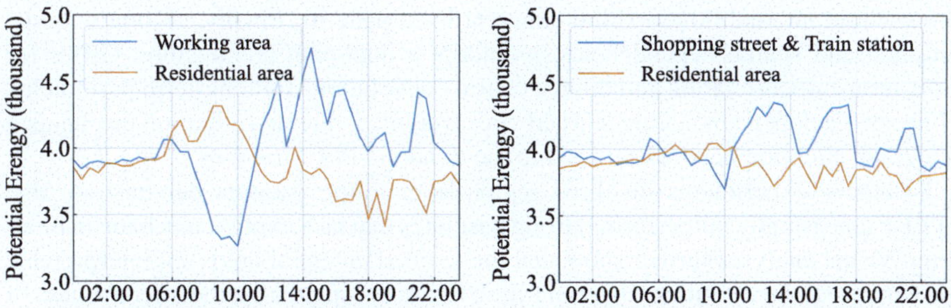

(a) Weekday potential dynamics

(b) Weekend potential dynamics

Fig. 3.11 Temporal dynamics of potentials on weekdays and weekends

3.5.4.3 Inter-Community Potential-Field Analysis

We next analyze and visualize inter-community flows, as shown in Fig. 3.10e and f. For each node, we compute its *total inter-community potential* as the sum of its intra-community and inter-community PEFs, and highlight the top 4% and bottom 4% of nodes in pink and yellow, respectively.

During the morning peak, a typical low-potential area appears in the northwestern part of Beijing—the Zhongguancun Science Park ("China's Silicon Valley")—which attracts substantial inter-community commuting from the southern residential regions. This suggests that rapidly developing southern neighborhoods provide major commuter flows toward the northern technology parks.

In the evening (Fig. 3.10f), the pattern reverses: southern residential areas become low-potential regions, drawing commuters back from the northern technology clusters. These findings indicate that the tidal potential effect extends beyond short-range intra-community commuting and also characterizes long-distance inter-community commuting, further validating the utility of PEFs for modeling the dynamic evolution of urban traffic.

3.6 Summary

This chapter presents a spatio-temporal potential-field model (**STPEF**), grounded in a gravity-inspired formulation, to capture the intrinsic mechanisms of region-level urban traffic flows.

To more precisely characterize human mobility, we design a discrete-time PEF extraction algorithm that decomposes the regional traffic-flow graph into multiple polytree-based subgraphs, each defining a potential energy field in which flows are constrained to move from higher- to lower-potential nodes. To further enhance predictive accuracy, we introduce a global-flow-aware community partitioning method that incorporates long-horizon temporal and spatial information, yielding communities whose internal flows exhibit coherent local patterns. Finally, to jointly model temporal and spatial dependencies, we adopt a dual-path architecture: GRUs for short-term autocorrelation and dilated causal convolutions for long-term periodicity, while spatial dependencies are learned via a weighted, directed attention mechanism in a graph attention network tailored to the dynamic PEF structure.

Extensive experiments on three region-level traffic datasets demonstrate that **STPEF** consistently outperforms strong baselines that lack explicit mechanism modeling. Visual analyses further show that the learned potential fields are interpretable and can reveal the evolution of urban traffic flows, substantiating the effectiveness of the potential-field perspective for uncovering internal mechanisms in spatio-temporal data.

Our approach assumes that observations are approximately stationary within each discrete time slice and that flows are defined on a grid, making it particularly suitable for structured, regular scenarios such as long-term urban traffic forecasting and analysis. In highly uncertain and rapidly changing environments, additional exogenous inputs (e.g. events, policies, or weather conditions) may need to be incorporated to maintain predictive accuracy.

References

1. Bai, L., Yao, L., Li, C., Wang, X., Wang, C.: Adaptive graph convolutional recurrent network for traffic forecasting. In: Proceedings of the 34th International Conference on Neural Information Processing Systems (NeurIPS'20), pp. 17804–17815 (2020)
2. Carey, H.C.: Principles of social science. JB Lippincott and Company (1871)
3. Chen, L., Zhang, D., Wang, L., Yang, D., Ma, X., Li, S., Wu, Z., Pan, G., Nguyen, T.M.T., Jakubowicz, J.: Dynamic cluster-based over-demand prediction in bike sharing systems. In: Proceedings of the 2016 ACM International Joint Conference on Pervasive and Ubiquitous Computing (UbiComp'16), pp. 841–852 (2016)
4. Cho, K., van Merriënboer, B., Gulcehre, C., Bahdanau, D., Bougares, F., Schwenk, H., Bengio, Y.: Learning phrase representations using RNN encoder–decoder for statistical machine translation. In: Proceedings of the 2014 Conference on Empirical Methods in Natural Language Processing, pp. 1724–1734 (2014)
5. Ji, J., Wang, J., Jiang, Z., Ma, J., Zhang, H.: Interpretable spatiotemporal deep learning model for traffic flow prediction based on potential energy fields. In: Proceedings of the 2020 IEEE International Conference on Data Mining (ICDM'20), pp. 1076–1081 (2020)
6. Jiang, Z., Xie, M., Sainju, A.M.: Geographical hidden Markov tree. IEEE Transactions on Knowledge and Data Engineering (TKDE) **33**(2), 506–520 (2021)
7. Kingma, D.P., Ba, J.: Adam: A method for stochastic optimization. In: Proceedings of the 3rd International Conference on Learning Representations (ICLR'15), pp. 1–15 (2015)
8. Kruskal, J.B.: On the shortest spanning subtree of a graph and the traveling salesman problem. Proceedings of the American Mathematical Society **7**(1), 48–50 (1956)
9. Li, M., Zhu, Z.: Spatial-temporal fusion graph neural networks for traffic flow forecasting. In: Proceedings of the 35th AAAI Conference on Artificial Intelligence (AAAI'21), pp. 4189–4196 (2021)
10. Liu, Y., Wang, F., Xiao, Y., Gao, S.: Urban land uses and traffic 'source-sink areas': Evidence from gps-enabled taxi data in shanghai. Landscape and Urban Planning **106**(1), 73–87 (2012)
11. Mazzoli, M., Molas, A., Bassolas, A., Lenormand, M., Colet, P., Ramasco, J.J.: Field theory for recurrent mobility. Nature Communications **10**(1), 3895 (2019)

12. Nash-Williams, C.S.J.: Edge-disjoint spanning trees of finite graphs. Journal of the London Mathematical Society **1**(1), 445–450 (1961)
13. Newman, M.E., Girvan, M.: Finding and evaluating community structure in networks. Physical Review E **69**(2), 026113 (2004)
14. Rebane, G., Pearl, J.: The recovery of causal polytrees from statistical data. In: Proceedings of the 3rd Conference on Uncertainty in Artificial Intelligence, p. 222–228 (1987)
15. Roy, J.R., Thill, J.C.: Spatial Interaction Modelling. Springer (2004)
16. Sainju, A.M., He, W., Jiang, Z.: A hidden Markov contour tree model for spatial structured prediction. IEEE Transactions on Knowledge and Data Engineering (TKDE) **34**(4), 1530–1543 (2020)
17. Shaojie Bai J. Zico Kolter, V.K.: Convolutional sequence modeling revisited. In: Proceedings of the 6th International Conference on Learning Representations (ICLR'18), pp. 1–20 (2018)
18. Sutskever, I., Vinyals, O., Le, Q.V.: Sequence to sequence learning with neural networks. In: Proceedings of the 28th International Conference on Neural Information Processing Systems (NeurIPS'14), pp. 3104–3112 (2014)
19. Taieb, S.B., Atiya, A.F.: A bias and variance analysis for multistep-ahead time series forecasting. IEEE Transactions on Neural Networks and Learning Systems **27**(1), 62–76 (2015)
20. Tian, C., Chan, W.K.: Spatial-temporal attention WaveNet: A deep learning framework for traffic prediction considering spatial-temporal dependencies. IET Intelligent Transport Systems **15**(4), 549–561 (2021)
21. Velickovic, P., Cucurull, G., Casanova, A., Romero, A., Liò, P., Bengio, Y.: Graph attention networks. In: Proceedings of the 6th International Conference on Learning Representations (ICLR'18), pp. 1–12 (2018)
22. Wang, J., Ji, J., Jiang, Z., Sun, L.: Traffic flow prediction based on spatiotemporal potential energy fields. IEEE Transactions on Knowledge and Data Engineering (TKDE) **35**(9), 9073–9087 (2023)
23. Wang, J., Jiang, J., Jiang, W., Li, C., Zhao, W.X.: LibCity: An open library for traffic prediction. In: Proceedings of the 29th ACM SIGSPATIAL International Conference on Advances in Geographic Information Systems, pp. 145–148 (2021)
24. Wang, J., Wu, N., Lu, X., Zhao, W.X., Feng, K.: Deep trajectory recovery with fine-grained calibration using Kalman filter. IEEE Transactions on Knowledge and Data Engineering (TKDE) **33**(3), 921–934 (2019)
25. Wang, J., Wu, N., Zhao, W.X.: Personalized route recommendation with neural network enhanced search algorithm. IEEE Transactions on Knowledge and Data Engineering (TKDE) **34**(12), 5910–5924 (2021)
26. Wang, L., Chai, D., Liu, X., Chen, L., Chen, K.: Exploring the generalizability of spatio-temporal traffic prediction: Meta-modeling and an analytic framework. IEEE Transactions on Knowledge and Data Engineering (TKDE) **35**(4), 3870–3884 (2023)
27. Wu, N., Wang, J., Zhao, W.X., Jin, Y.: Learning to effectively estimate the travel time for fastest route recommendation. In: Proceedings of the 28th ACM International Conference on Information and Knowledge Management (CIKM'19), pp. 1923–1932 (2019)
28. Wu, T., Tang, L., Chen, H., Wang, Z., Qiu, Q.: Application of source-sink landscape influence values to commuter traffic: A case study of xiamen island. Sustainability **9**(12), 2366 (2017)
29. Yao, H., Tang, X., Wei, H., Zheng, G., Li, Z.: Revisiting spatial-temporal similarity: A deep learning framework for traffic prediction. In: Proceedings of the 33rd AAAI Conference on Artificial Intelligence (AAAI'19), pp. 5668–5675 (2019)
30. Yu, B., Yin, H., Zhu, Z.: Spatio-temporal graph convolutional networks: A deep learning framework for traffic forecasting. In: Proceedings of the 27th International Joint Conference on Artificial Intelligence (IJCAI'18), pp. 3634–3640 (2018)
31. Zhang, J., Zheng, Y., Qi, D., Li, R., Yi, X.: DNN-based prediction model for spatio-temporal data. In: Proceedings of the 24th ACM SIGSPATIAL International Conference on Advances in Geographic Information Systems, pp. 1–4 (2016)
32. Zhang, J., Zheng, Y., Qi, D., Li, R., Yi, X., Li, T.: Predicting citywide crowd flows using deep spatio-temporal residual networks. Artificial Intelligence **259**(1), 147–166 (2018)

Chapter 4
Differential Equation Network

Abstract Road network-level traffic prediction focuses on the accurate prediction of traffic conditions (such as traffic flow) on specific roads and intersections, which is crucial for intelligent traffic management and optimization. Compared with the region-level prediction in the previous chapter, it can provide information more consistent with road structures and support applications such as intelligent signal control, path planning, and dynamic tolling. However, existing road network traffic prediction models typically assume that traffic states are generated in discrete time slots and remain constant throughout each time slot, which prevents these models from accurately capturing the continuous-time evolution process of traffic flow. Therefore, this chapter introduces a Spatio-Temporal Differential Equation Network (STDEN) based on the continuity equation to address this problem.

4.1 Modeling Motivation

Traffic flow prediction over urban road networks plays a crucial role in modern urban traffic management. With the continuous increase in urban traffic volume, congestion problems have become increasingly severe, leading to uncertainty in travel time, frequent traffic accidents, and intensified environmental pollution. Therefore, accurate road network traffic flow prediction not only helps individual users with travel planning but also provides a basis for traffic managers to adjust and optimize traffic signals in real time and make decisions on road construction and planning [12].

Through the prediction of urban road network traffic flow, traffic management departments can identify potential bottleneck sections, predict possible congestion situations, and timely implement traffic diversion measures, thereby effectively reducing traffic accidents and congestion. In addition, road network traffic prediction also provides support for intelligent transportation systems and autonomous driving technologies, helping these technologies make rational decisions in complex urban traffic environments and avoid traffic conflicts and accidents. For urban residents, accurate traffic flow prediction can help them choose the optimal travel time and route, reduce unnecessary time waste and waiting, and improve travel efficiency.

J. Wang and J. Ji, *Mechanism-Driven Explainable Urban Spatio-Temporal Prediction*, Big Data Management, https://doi.org/10.1007/978-981-92-0662-9_4

At the macro level, accurate road network traffic prediction can also support the construction of sustainable urban transportation systems. Through rational traffic guidance and scheduling, it reduces energy consumption and pollution emissions, laying a foundation for achieving green transportation goals.

Unlike the grid-based traffic prediction methods in Chap. 3, urban road network traffic flow prediction focuses on analyzing the traffic flow dynamics between specific roads and intersections in the road network. Grid-based methods perform traffic analysis by dividing the city into regular grid cells, where each cell represents a region. However, urban road network traffic flow prediction focuses on specific roads and intersections in the road network, and can provide information that is highly consistent with the road structure. When traffic data is defined on a non-Euclidean structure such as a road network, standard convolutional neural networks for grid-based regional traffic flow prediction struggle to effectively capture spatial dependencies on non-Euclidean structures due to their design limitations on regular grid structures. In contrast, graph neural networks are a more suitable approach, as their architecture is naturally adapted to process non-Euclidean structure data and can efficiently capture spatial dependencies on such structures [13, 19].

Although existing road network traffic flow prediction models have achieved certain performance on multiple benchmark datasets, most of these models are based on the discrete-time assumption, which partitions time into equally spaced time slices and assumes that the traffic state within each slice remains constant. For instance, if the time interval adopted by such methods is 10 min, they will compute the average value of a given road segment over the 10 min period as the traffic data observation for the current time slice, and assume that the traffic state remains unchanged within this 10 min window. While this assumption simplifies model construction, it neglects the inherent continuous evolutionary properties of traffic flow, making existing methods unable to accurately capture the small-time-scale dynamic changes of traffic flow within time slices. Specifically, road network traffic flow prediction methods based on the discrete-time assumption have the following two limitations.

- First, there exists a contradiction between the discrete-time assumption and the continuity of traffic flow. Particularly when considering the energy continuity principle of the road network potential energy field, the model struggles to truly reflect the dynamic evolution of traffic flow. The mechanism model of the road network potential energy field describes the variations of traffic flow over continuous time scales, yet the discrete-time assumption limits the model's ability to capture such continuous dynamics, thereby reducing its prediction accuracy for complex traffic flow characteristics.
- Second, under the discrete-time assumption, existing methods primarily employ deep learning models with discrete hidden layers to model the traffic state within each time slice. Such models typically fix the number of hidden layers in their network structure, meaning that once the model is designed, the number and structure of hidden layers cannot be dynamically adjusted. However, in continuous-time modeling, the number of hidden states should be time-dependent—i.e., as time elapses, the number and structure of hidden states ought to change dynamically.

For the continuous-time modeling of traffic flow, the model needs to flexibly adjust the number of hidden layers to adapt to traffic state changes across different time periods.

Consequently, existing deep learning models based on discrete hidden layers cannot be directly applied to the modeling of continuous-time data.

To address the mismatch between existing discrete modeling methods and the requirements of continuous traffic flow prediction, this chapter proposes a Spatio-Temporal Differential Equation Network (STDEN) model based on the continuity Eq. [9]. This model embeds the continuity of time into its structure by introducing the energy continuity equation followed by the potential energy field as a constraint. Specifically, the model characterizes the continuous spatial and temporal changes of traffic flow by establishing a potential energy field differential equation closely related to traffic flow evolution, which enables the model to follow the spatial evolution trajectory of the potential energy field on the road network over continuous time during the prediction process.

First, to overcome the limitations of the discrete time assumption in traditional traffic flow prediction, this chapter introduces a road network potential energy field with continuous time characteristics, and constructs a continuity differential equation that describes the dynamic evolution of the road network potential energy field based on the energy continuity equation, so as to capture the continuous variation characteristics of traffic flow in the temporal and spatial dimensions.

Specifically, in physics, the continuity equation is a mathematical formulation of the law of conservation, which is widely used to characterize the spatio-temporal transmission laws of physical quantities such as energy, mass, and momentum [15]. Among them, due to the flexibility of its theoretical framework, the energy continuity equation can be used to describe the transmission and distribution of energy in different physical scenarios, and its core idea is that there exists a dynamic equilibrium relationship between the temporal rate of change of a local physical quantity and its transport quantity through the boundary. This characteristic is highly similar to the spatio-temporal evolution mechanism of traffic flow driven by the road network potential energy field, and both reflect certain conservation constraints and continuity characteristics.

Therefore, based on the theoretical foundation of the continuity equation for energy transmission, this chapter deduces the continuity differential equation describing the evolution of the road network potential energy field by analogy with the evolution mechanism of road network traffic flow under the constraints of the potential energy field. This mathematical model can not only accurately characterize the dynamic transmission process of traffic flow in the road network but also provide solid theoretical support for constructing traffic flow prediction models with continuous time modeling capabilities.

Second, to address the issue that traditional neural networks with discrete hidden layers cannot represent a dynamic number of hidden states, under the guidance of the potential energy field differential equation, a spatio-temporal differential equation network with continuous hidden layers is designed to enhance the model's ability to model the evolution of traffic flow over continuous time.

Specifically, the neural differential equation network [4] proposes the idea of continuousizing the transfer process between neural network layers, transforming the recursive update of traditional discrete layers into an initial value problem of differential equations. Under this framework, the number of network layers is no longer discrete and fixed; instead, hidden states are generated by solving differential equations over continuous time. This chapter extends the conventional differential equation network. Under the guidance of the potential energy field differential equation, we replace the differential operator with the graph Laplacian and design a spatio-temporal differential equation network with continuous hidden layers based on the road network structure. The overall framework of STDEN consists of an encoder that maps traffic flow to the potential energy field, a differential equation network that continuously predicts the dynamics of the potential energy field over time, and a decoder that generates traffic flow predictions from the potential energy field.

This model can naturally embed the evolution laws of the road network potential energy field over continuous time, thereby achieving accurate modeling of temporal dynamic changes in traffic flow prediction tasks. Compared with traditional discrete-time methods, this model not only has stronger generalization capability for temporal continuity but also ensures that the prediction results comply with the mechanistic constraints of the potential energy field, thereby improving the interpretability and generalizability of the model.

4.2 Basic Concepts and Problem Definition

This subsection mainly introduces the basic concepts involved in this chapter, including road networks, road network traffic flow, road network potential energy field, traffic flow sequences, and potential energy field sequences. Then, the problem definition of road network-based traffic flow prediction is presented.

Definition 4.1 (*Road Network*) A road network can be represented as a directed graph $\mathcal{G} = (\mathcal{V}, \mathcal{E}, W)$, where $\mathcal{V} = \{v_1, \ldots, v_n\}$ denotes the set of nodes, $\mathcal{E} \subseteq \mathcal{V} \times \mathcal{V}$ is the set of edges, and $W \in \mathbb{R}^{n \times n}$ is the weighted adjacency matrix. A node $v_i \in \mathcal{V}$ represents a road intersection or a road endpoint, while an edge $e_{ij} \in \mathcal{E}$ denotes a directed road from node v_i to node v_j.

Definition 4.2 (*Road Network Traffic Flow*) Road network traffic flow is defined as a feature of edges in a road network. For a given road network $\mathcal{G}$, the total traffic flow of edge e_{ij} within a specified time period is denoted as f_{ij}. Correspondingly, the traffic flow of the entire road network within this time period can be represented as a vector $\boldsymbol{f} = (f_{ij}) \in \mathbb{R}^{|\mathcal{E}|}$, where $|\mathcal{E}|$ is the size of the edge set.

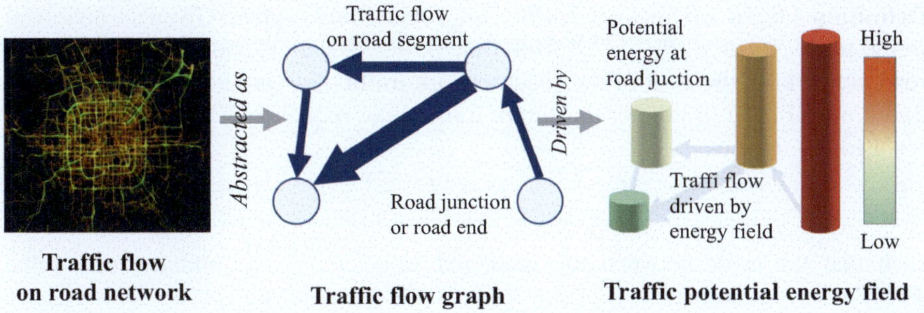

Fig. 4.1 Traffic flow and potential energy field on a road network

Definition 4.3 (*Road Network Potential Energy Field*) For each node i of the road network $\mathcal{G}$, we define its network potential energy as $z_i \in \mathbb{R}$. The traffic potential energy field of the entire road network can be represented as the vector $z = (z_1, \ldots, z_n)^\top \in \mathbb{R}^n$.

As shown in Fig. 4.1, the potential energy field on the road network can be regarded as the underlying driving force of traffic flow, similar to the driving effect of the gravitational field on water flow. The traffic flow on the corresponding edge can be expressed as the potential energy gradient between adjacent nodes, i.e.,

$$f_{ij} = -(\nabla z)_{ij} = -(z_i - z_j), \tag{4.1}$$

where ∇ denotes the graph gradient operator, and the negative sign is used to indicate the direction of flow. According to Eq. (4.1), the traffic flow in the road network can be understood as an "energy transmission" process of the traffic potential energy field, and this viewpoint forms the basic theoretical foundation of the proposed model.

Definition 4.4 (*Road Network Traffic Flow Sequence and Potential Energy Field Sequence*) Time is divided into regular time slices. Let $\boldsymbol{f}^{(t)}$ denote the traffic flow observed in time slice $t \in \mathbb{N}$; then the traffic flow sequence can be expressed as the time series

$$\boldsymbol{F}^{(0:t)} = \left(\boldsymbol{f}^{(0)}, \ldots, \boldsymbol{f}^{(t)}\right), \tag{4.2}$$

Similarly, the potential energy field sequence, which represents the temporal evolution process of the Potential Energy Field (PEF), can be expressed as

$$\mathbf{Z}^{(0:t)} = \left(z^{(0)}, \ldots, z^{(t)}\right). \tag{4.3}$$

Based on the above basic concepts, the problem of road network traffic flow prediction can be defined as follows.

Definition 4.5 (*Road Network Traffic Flow Prediction Problem*) Given the historical traffic flow sequence $\boldsymbol{F}^{(t-T+1:t)}$ from time $t-T+1$ to t, and the corresponding road network $\mathcal{G}$, the goal of the road network traffic flow prediction problem is to find a model $m(\cdot)$ to predict the future traffic flow sequence $\boldsymbol{F}^{(t+1:t+H)}$ from time $t+1$ to $t+H$, i.e.,

$$\hat{\boldsymbol{F}}^{(t+1:t+H)} = m\left(\boldsymbol{F}^{(t-T+1:t)}\right), \tag{4.4}$$

such that the error between the predicted value $\hat{\boldsymbol{F}}^{(t+1:t+H)}$ and the true value $\boldsymbol{F}^{(t+1:t+H)}$ is minimized.

4.3 Physics-Guided Road Network Traffic Flow Modeling

To address the mismatch between existing discrete modeling methods and the demand for continuous traffic flow prediction, this chapter attempts to construct a physical model of traffic flow in road networks. Based on the potential energy field concept proposed in the previous chapter, we introduce the continuity equation of energy and derive the energy continuity differential equation of the potential energy field to simulate the continuous evolution of traffic flow. Furthermore, taking this differential equation as a constraint condition, we embed temporal continuity into the structure of the deep learning model and construct a physics-guided deep Spatio-Temporal Differential Equation Network (STDEN) to learn the continuous temporal changes of traffic flow.

4.3.1 Model Framework

Classic data-driven methods usually directly model the correlation between $\boldsymbol{F}^{(t-T+1:t)}$ and $\boldsymbol{F}^{(t+1:t+H)}$ to generate traffic flow prediction results. This chapter innovatively proposes a physics-guided traffic flow prediction framework (see Fig. refSTDEN:fig:framework), which consists of two parts: an explicit deep learning model Spatio-Temporal Differential Equation Network (STDEN) and an implicit physical dynamic process Physics Guided Flow Modeling (PGFM).

In PGFM, the physics-based continuity equation is used to describe the conversion relationship between the Potential Energy Field (PEF) and traffic flow, and this process is denoted as PF-Trans in Fig. 4.2. This subsection will present a detailed introduction to the implicit PGFM.

In the implicit physical dynamic process, a deep neural network model STDEN is constructed, which consists of the following three core components:

- An RNN-based encoder, which encodes traffic flow sequences into the initial state of the PEF;

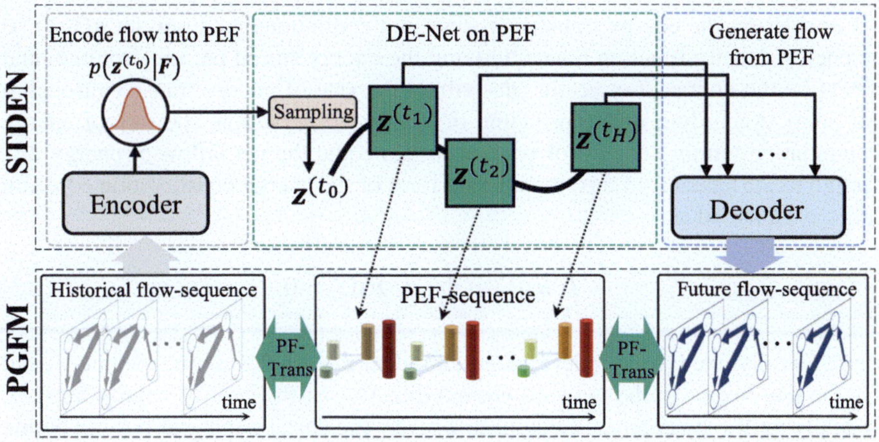

Fig. 4.2 Overview of the physics-guided deep spatio-temporal differential equation network structure

- A Differential Equation (DE) network, a deep learning model designed under the guidance of the evolution mechanism of the PEF, which predicts the spatio-temporal evolution trajectory of the PEF;
- A decoder constrained by Eq. (4.1), which generates road network traffic flow based on the predicted PEF.

It should be emphasized that each layer structure of the DE network corresponds to the potential energy dynamic evolution mechanism in PGFM. The specific details of STDEN will be discussed in Sect. 4.4.

This model unifies physical dynamic processes and deep learning models under the same framework, thus is called a *Physics-Guided Neural Network(PGNN)*.

4.3.2 Physics-Based Continuity Equation

This section adopts the continuity equation to model the relationship between traffic flow sequences and PEF sequences. First, a brief introduction to the basic principles of the continuity equation is provided.

In physics, the continuity equation, as a mathematical formulation of the conservation law, is widely used to describe the spatio-temporal transport laws of extensive quantities such as mass, energy, momentum, and charge [15]. This equation essentially reflects the dynamic equilibrium relationship between the time rate of change of local physical quantities and the transport quantity through the boundary. Here, we focus on discussing its specific application forms and physical connotations in the problem of energy transport.

Considering the energy transfer process in a continuous medium, let $\boldsymbol{u}(x, t)$ be the energy density function (characterizing the energy stored per unit volume) and $\boldsymbol{q}(x, t)$ be the energy flux vector (describing the rate of energy transfer through a unit area). According to the principle of energy conservation, the rate of energy accumulation within the control volume should equal the net inflow of energy flux through its surface. From this, the integral form of the energy conservation equation can be established as:

$$\frac{\partial}{\partial t}\int_V \boldsymbol{u}\,\mathrm{d}V + \oint_{\partial V} \boldsymbol{q}\cdot\boldsymbol{n}\,\mathrm{d}S = 0, \tag{4.5}$$

where S is any fixed (time-independent) closed surface enclosing the volume V, and $\boldsymbol{n}$ is the unit vector in the outward normal direction of the control volume surface. By applying the divergence theorem to convert the surface integral into a volume integral and considering the arbitrariness of the control volume, the differential form of the energy continuity equation can be derived as:

$$\frac{\partial \boldsymbol{u}}{\partial t} + \mathrm{div}\boldsymbol{q} = 0, \tag{4.6}$$

where div is the divergence operator. For a three-dimensional Cartesian coordinate system, its definition is given by

$$\mathrm{div}\boldsymbol{q} = \left(\frac{\partial}{\partial x_1}, \frac{\partial}{\partial x_2}, \frac{\partial}{\partial x_3}\right)\cdot\left(q_{x_1}, q_{x_2}, q_{x_3}\right) = \frac{\partial q_{x_1}}{\partial x_1} + \frac{\partial q_{x_2}}{\partial x_2} + \frac{\partial q_{x_3}}{\partial x_3}, \tag{4.7}$$

The first term in Eq. (4.6) characterizes the instantaneous rate of change of energy density, while the second term describes the spatial divergence of the energy flux. The property that their signs are opposite indicates that when energy converges to a certain region ($\mathrm{div}\boldsymbol{q} < 0$), the energy density of that region will increase over time; conversely, it will decrease. This differential form of the conservation law overcomes the macroscopic limitations of the integral form and can finely characterize the local dynamic behavior of the energy field [2].

It is worth noting that Eq. (4.6) exhibits the characteristic of a universal framework, and its specific physical meaning needs to be further clarified in conjunction with constitutive relations. For example, in Fourier's law of thermal conduction,[1] the energy flux $\boldsymbol{q}$ and temperature gradient satisfy $\boldsymbol{q} = -\kappa\nabla T$, where κ is the thermal conductivity and T is the temperature; while in electromagnetic field energy transmission, $\boldsymbol{q}$ corresponds to the Poynting vector [8].

The flexibility of this theoretical framework allows the continuity equation to adapt to the modeling requirements of energy transmission in different physical scenarios.

[1] Fourier's law of thermal conduction: https://en.wikipedia.org/wiki/Thermal_conduction.

From the above analysis, it can be seen that the continuity equation not only provides a rigorous framework for the mathematical description of energy transfer, but also establishes a cross-scale analytical system for energy dynamics through coupling with other physical laws (e.g., thermodynamic constitutive relations, Maxwell's equations, etc.). This modeling approach based on conservation laws has become a fundamental paradigm in fields such as continuum mechanics, heat transfer, and electromagnetic field theory.

It is worth noting that the differential equation in Eq. (4.6) establishes the relationship between the energy volume and the 'transmission' of this energy, that is, the change in energy density leads to energy transmission in space. For example, heat always transfers from hot areas to cold areas. This shares a similar physical mechanism with the continuity equation of the traffic PEF in the next subsection.

4.3.3 Continuity Equation of Traffic Potential Energy Field

In road networks, the traffic flow driven by the traffic PEF is essentially a specific instance of the continuity equation. The traffic PEF can be analogized to the energy field in the continuity equation, where the potential energy per unit volume corresponds to the energy density, and the traffic flow corresponds to the energy flux.

By comparing the relationship between potential energy-driven traffic flow and the energy continuity equation, it can be found that traffic flow in road networks can be regarded as an energy transmission process in a PEF. Based on this analogy, this section proposes a continuity equation for the traffic PEF to describe the dynamic evolution relationship between traffic flow sequences and PEF sequences in road networks. Next, the analogy process is first introduced, and then the continuity equation for the traffic PEF is presented.

4.3.3.1 Energy Density and Traffic Potential Energy

In a given road network $\mathcal{G} = \{\mathcal{V}, \mathcal{E}, W\}$, for each node $v_i \in \mathcal{V}$, its energy density is defined as u_i. The potential energy of a node is proportional to its energy density, i.e.,

$$z_i = \phi_i \cdot u_i, \tag{4.8}$$

where ϕ_i is a trainable node volume. This parameter is determined by the endogenous features of the node and is analogous to mass in a gravitational field.

For the entire node set $\mathcal{V}$, the relationship between the potential energy vector $z = (z_1, \ldots, z_n)^\top$ and the energy density vector $\boldsymbol{u} = (u_1, \ldots, u_n)^\top$ can be expressed as

$$z = \boldsymbol{\phi} \odot \boldsymbol{u}, \tag{4.9}$$

where $\boldsymbol{\phi} = (\phi_1, \ldots, \phi_n)^\top$ and the symbol $\odot$ denotes the Hadamard product (element-wise multiplication).

4.3.3.2 Energy Flux and Traffic Flow

In the traffic system defined on $\mathcal{G}$, the transmission of traffic potential energy is limited to the edge set $\mathcal{E}$, that is, the potential energy can only propagate along the edges (road segments) in the road network. Therefore, the energy flux of the edge $e_{ij} \in \mathcal{E}$ is defined as q_{ij}.

The traffic flow f_{ij} can be regarded as the main component of the energy flux q_{ij}. To describe the relationship between the two, a shared parameter α is introduced, which is used to measure the contribution rate of traffic flow to energy flux on each edge, thus establishing the following relationship:

$$\boldsymbol{f} = \alpha^{-1}\boldsymbol{q}, \tag{4.10}$$

where $\boldsymbol{f} = (f_{ij})^\top$ denotes the traffic flow vector and $\boldsymbol{q} = (q_{ij})^\top$ denotes the energy flux vector.

4.3.3.3 Differential Equation of Potential Energy Field

Substituting Eqs. (4.9) and (4.10) into the energy continuity Eq. (4.6) and performing simple derivations yields

$$\boldsymbol{\phi}^{-1} \odot \frac{\partial z}{\partial t} + \alpha \mathrm{div} \boldsymbol{f} = 0, \tag{4.11}$$

This equation is referred to as the continuity equation for energy transmission in the traffic PEF.

According to Eq. (4.1), the traffic flow on edges in a road network can be expressed as the gradient of the potential energy of adjacent nodes, i.e., $\boldsymbol{f} = -\nabla z$. Substituting this into Eq. (4.10) to replace the traffic flow $\boldsymbol{f}$, we obtain

$$\boldsymbol{\phi}^{-1} \odot \frac{\partial z}{\partial t} - \alpha \mathrm{div} \nabla z = 0. \tag{4.12}$$

This function is a differential equation (DE) with respect to the potential energy z, so it is referred to as the *potential energy differential equation.*

Since the divergence of the gradient defines the Graph Laplacian Δ [3], whose form is given by

$$\Delta = -\mathrm{div} \nabla, \tag{4.13}$$

thus the differential equation of the PEF can be transformed into

$$\frac{\partial z}{\partial t} = -\boldsymbol{\phi} \odot (\alpha \Delta z) . \tag{4.14}$$

This equation describes the temporal evolution process of the PEF z, where $\boldsymbol{\phi}$ reflects the volume property of nodes, and α controls the influence of the potential energy gradient on potential energy changes. The equation indicates that the rate of potential energy change is regulated by the Graph Laplacian Δ, revealing the mathematical relationship between potential energy diffusion and traffic flow.

4.3.3.4 Physical Interpretation of the Potential Energy Field Differential Equation

Equation (4.14) represents the evolutionary dynamic equation of the traffic PEF sequence $\mathbf{Z} = \left(z^{(t)}\right)$. From the perspective of discretization, this equation can be expressed as

$$z^{(t+1)} = z^{(t)} - \boldsymbol{\phi} \odot \left(\alpha \Delta z^{(t)}\right) . \tag{4.15}$$

It indicates that the PEF z at each time step t depends on its state at the previous moment, while being influenced by the diffusion effect under the action of the graph Laplacian operator Δ, which is a spatio-temporal evolution process.

This dynamic equation is analogous to the energy diffusion process in physical systems, where energy propagates along the network structure and drives the dynamic changes of the overall traffic flow.

4.4 Spatio-Temporal Differential Equation Network

In Eq. (4.14), the evolution of the PEF sequence is expressed in the form of a potential energy field differential equation. To enhance the modeling capability of this differential equation for real-world traffic data, this section designs and implements a neural network model based on this differential equation.

This framework combines data-driven methods with mechanistic modeling, ensuring that the model can not only effectively capture complex data patterns but also adhere to the laws of physical dynamics. The overall structure of the framework is shown in the upper part of Fig. 4.2.

4.4.1 Differential Equation Networks

The method in this chapter is based on the Neural Ordinary Differential Equation Network (DE-Net) [4], whose core idea is to render continuous the discrete inter-layer propagation process in traditional deep neural networks and transform it into an initial value problem of ordinary differential equations.

To clarify this theoretical framework, this section first reviews the basic structure of the Residual Network (ResNet), and then derives its continuous form through mathematical limits.

4.4.1.1 Discrete Recursive Structure of Residual Networks

The standard Residual Network [7] alleviates the vanishing gradient problem of deep networks by introducing skip connections. The state update rule for its t-th layer can be expressed as

$$\boldsymbol{h}_{t+1} = \boldsymbol{h}_t + \mathcal{F}(\boldsymbol{h}_t, \theta_t), \tag{4.16}$$

where $\boldsymbol{h}_t$ is the hidden state of the t-th layer,[2] $\mathcal{F}(\cdot)$ is a parameterized residual function (usually composed of fully connected layers or convolutional layers), and θ_t is the trainable network parameter of the t-th layer.

In traditional ResNet, the parameter θ_t of each layer is independent, leading to a linear growth of the network parameter count with depth; whereas DE-Net treats the layer index t as a continuous time variable and significantly reduces the model complexity through a parameter sharing mechanism.

4.4.1.2 Derivation of Continuousization for Discrete Recurrence

To establish a connection with differential equations, rewrite Eq. (4.16) into the form of a difference equation. Define the time step for state update as l; then we have

$$\frac{\boldsymbol{h}_{t+l} - \boldsymbol{h}_t}{l} = \frac{\boldsymbol{h}_{t+l} - \boldsymbol{h}_t}{(t+l) - t} = \mathcal{F}(\boldsymbol{h}_t, \theta_t), \tag{4.17}$$

When $l = 1$, the above equation degenerates to the original residual formula.

If we map the discrete points of network depth $\{t, t+1, \ldots, t+T\}$ to the continuous time interval $[t, t+T]$ and let the step size tend to zero, i.e., $l \to 0$, then the update process of state $\boldsymbol{h}$ in the difference Eq. (4.17) becomes a differential equation:

$$\frac{\partial \boldsymbol{h}(t)}{\partial t} = \mathcal{F}\left(\boldsymbol{h}(t), t, \theta\right), \tag{4.18}$$

[2] In the model of this chapter, the layers of DE-Net are only related to the time step t, so Eq. (4.16) uses t as the index for the network layer number.

Table 4.1 Comparison of characteristics between ResNet and DE-Net

Characteristic	ResNet	DE-Net
Definition of depth	Fixed number of discrete layers	Adjustable continuous time T
Parameter sharing	Independent parameters for each layer	Globally shared parameter θ
Memory consumption	Linear growth with the number of layers	Constant
Dynamic flexibility	Fixed discrete transitions	Continuously differentiable trajectory
Physical integration capability	Limited	Can embed priors via constraint $\mathcal{F}$

At this point, the hidden state $\boldsymbol{h}(t)$ becomes a continuous function of time $t \in (0, T]$. The residual function $\mathcal{F}$ is transformed into a dynamic function parameterized by a neural network, which governs the evolution process of the dynamic system [16, 18]. The parameter θ is shared across the time dimension, achieving parameter efficiency for the continuous-depth model.

The function $\mathcal{F}(\cdot, \theta)$ here is a neural network, which provides powerful representation learning capabilities for complex real-world data. That is to say, Eq. (4.18) represents a DE-Net, whose characteristics are compared in detail with those of ResNet in Table 4.1.

4.4.1.3 Mathematical Connotations of Continuous Dynamical Systems

Equation (4.18) essentially describes the evolution process of a dynamical system: the rate of change of the initial state $\boldsymbol{h}(0)$ (network input) with respect to time t is determined by $\mathcal{F}$. By solving this differential equation via numerical integration, the final state $\boldsymbol{h}(T)$ (network output) can be obtained as:

$$\boldsymbol{h}(T) = \boldsymbol{h}(0) + \int_0^T \mathcal{F}(\boldsymbol{h}(t), t, \theta)dt. \tag{4.19}$$

This process has the following characteristics:

- *Depth Continuity*: In traditional neural networks, the "depth" of the network is usually defined by a fixed number of layers, where each layer corresponds to specific parameters and computation processes. However, in this novel network architecture, the network depth is controlled by the length of the integration interval T rather than a fixed number of layers. This design allows the network depth to change continuously instead of being discrete as in traditional networks. By adjusting the integration endpoint, the model capacity can be flexibly expanded.

- *Parameter Sharing*: In traditional networks, each layer typically has its own independent parameters, which leads to rapid growth in the number of parameters. Especially in deep networks, the problem of parameter redundancy is particularly prominent. However, the dynamical function $\mathcal{F}(\cdot, \theta)$ shares the network parameters θ throughout the entire time domain, avoiding the redundancy of inter-layer parameters in traditional networks [18].
- *Physical Interpretability*: If $\mathcal{F}$ is combined with known physical laws (such as conservation of energy and conservation of momentum), a hybrid model integrating mechanistic modeling and data-driven approaches can be constructed [16], endowing the model with physical interpretability. This hybrid model can not only fully utilize the information in the data but also enhance the model's interpretability by leveraging known physical laws.

4.4.2 Traffic Potential Energy Field Differential Equation Network

This section applies the idea of DE-Net to the PEFs and designs a neural network model based on this differential equation for PEFs.

4.4.2.1 Potential Energy Field Differential Equation Network

Inspired by DE-Net, the potential energy field differential equation in Eq. (4.14) can be modeled as

$$\frac{\partial z^{(t)}}{\partial t} = \mathcal{F}_{\mathcal{G}}\left(\Phi, t, z^{(t)}\right), \tag{4.20}$$

where Φ denotes all trainable parameters, including $\boldsymbol{\phi}$ and α. The function $\mathcal{F}_{\mathcal{G}}$ is guided by the mechanistic model in Eq. (4.14) and modeled in the form of a neural network.

According to Eq. (4.14), the function $\mathcal{F}_{\mathcal{G}}$ can be expressed in the form of a residual graph convolutional network (GCN), where the repeated neural network layer is given by

$$\mathcal{F}_{\mathcal{G}}\left(\Phi, t, z^{(t)}\right) = -\boldsymbol{\phi} \odot \operatorname{Tanh}(\alpha \Delta z). \tag{4.21}$$

Here, Δ denotes the graph Laplacian, which is used to calculate the difference between the state z_i of node i and the states of its neighboring nodes. This process is equivalent to aggregating the states of nodes within the receptive field using α as the convolution kernel. In addition, $\operatorname{Tanh}(\cdot)$ is the hyperbolic tangent activation function, which endows the model with nonlinear modeling capability. Finally, the computed feature map is element-wise weighted using the weights provided by ϕ_i

(where $\odot$ denotes element-wise multiplication) to adjust the contributions of different nodes during the model learning process, thereby forming the output of the entire network.

From a spatial perspective, when t takes discrete values, Eq. (4.21) is equivalent to a residual GCN, where the input of each layer is the state $z^{(t)}$ of all nodes in the road network. From a temporal perspective, since t is a continuous variable, $z^{(t)}$ can be computed at any $t \in \mathbb{R}$. Based on this, the differential equation network defined by Eq. (4.21) is named the "spatio-temporal" differential equation network in this chapter to highlight *its ability to model continuity in both temporal and spatial dimensions*.

This modeling approach combines graph neural networks and neural differential equations. It can not only efficiently extract spatial dependencies but also flexibly adapt to time-series data with continuous or irregular time intervals, thus holding significant value for spatio-temporal data modeling tasks such as traffic flow prediction.

4.4.2.2 Potential Energy Field Encoder

In the proposed model, for a given initial state $z^{(t_0)}$, the PEF sequence $\boldsymbol{Z}^{(t_1:t_H)}$ can be computed using the neural ODE solver [4], i.e.,

$$\boldsymbol{Z}^{(t_1:t_H)} = \text{ODEsolver}\left(\mathcal{F}_{\mathcal{G}}, \Phi, z^{(t_0)}, [t_0, \ldots, t_H]\right), \tag{4.22}$$

where $z^{(t_0)}$ is sampled from the distribution $p\left(z^{(t_0)}\right)$.

The above distribution $p\left(z^{(t_0)}\right)$ is modeled by the conditional probability distribution of the historical traffic flow sequence $\boldsymbol{F}^{(t_0-T+1:t_0)}$, which takes the form of

$$p\left(z^{(t_0)} \middle| \boldsymbol{F}^{(t_0-T+1:t_0)}\right) = \mathcal{N}\left(\mu_{z^{(t_0)}}, \sigma_{z^{(t_0)}}\right), \tag{4.23}$$

where the mean $\mu_{z^{(t_0)}}$ and standard deviation $\sigma_{z^{(t_0)}}$ are extracted from the historical traffic flow sequence $\boldsymbol{F}^{(t_0-T+1:t_0)}$ using a Gated Recurrent Unit (GRU) [5], i.e.,

$$\left\{\mu_{z^{(t_0)}}, \sigma_{z^{(t_0)}}\right\} = g\left(\text{GRU}\left(\boldsymbol{F}^{(t_0-T+1:t_0)}\right)\right). \tag{4.24}$$

In this chapter, a GRU is employed as the encoder to extract key information from the historical traffic flow sequence $\boldsymbol{F}^{(t_0-T+1:t_0)}$. Here, $g(\cdot)$ is a fully connected network that maps the final hidden state of the GRU to the mean and standard deviation of the initial state $z^{(t_0)}$.

To implement a differentiable "sampling" operation, we introduce the reparameterization trick [10] into the generation process of $z^{(t_0)}$ in Eq. (4.23). Specifically, given a batch of training data, $z_i^{(t_0)}$ for each node i is calculated as

$$z_i^{(t_0)} = \mu_{z^{(t_0)}} + \epsilon_i \sigma_{z^{(t_0)}}, \tag{4.25}$$

where ϵ_i is sampled from the standard normal distribution $\mathcal{N}(0, 1)$. Through this method, $z_i^{(t_0)}$ corresponding to the training data remains fixed under the given conditions, thereby ensuring the differentiability of the backpropagation process in neural network training. In the prediction phase, for each input sample, ϵ_i is resampled to generate the corresponding $z_i^{(t_0)}$.

4.4.2.3 Potential Energy Field Decoder

Given the initial state $z^{(t_0)}$, the complete PEF sequence $\boldsymbol{Z}^{(t_1:t_H)}$ for future time steps can be computed through the process described in the previous subsection. Subsequently, by performing calculations using Eq. (4.1), the corresponding traffic flow sequence $\hat{\boldsymbol{F}}^{(t_1:t_H)}$ can be directly decoded from $\boldsymbol{Z}^{(t_1:t_H)}$ as the prediction of future traffic flow states. Note that this process is an exact calculation and does not require the introduction of additional parameters; if the PEF is predicted accurately enough, the traffic flow prediction will also be accurate.

The error between $\hat{\boldsymbol{F}}^{(t_1:t_H)}$ obtained through decoding and the ground-truth traffic flow sequence $\boldsymbol{F}^{(t_1:t_H)}$ can be calculated as

$$\mathcal{L} = \frac{1}{H|\mathcal{E}|} \sum_{\tau=t+1}^{t+H} \sum_{i=1}^{|\mathcal{E}|} \left| f_i^{(\tau)} - \hat{f}_i^{(\tau)} \right|, \tag{4.26}$$

where H is the prediction time horizon, $|\mathcal{E}|$ is the number of edges in the road network, and $f_i^{(\tau)} \in \boldsymbol{F}^{(t_1:t_H)}$ and $\hat{f}_i^{(\tau)} \in \hat{\boldsymbol{F}}^{(t_1:t_H)}$ denote the traffic flow of the i-th edge at time step τ in the ground-truth road network traffic flow and the predicted traffic flow, respectively. This error serves as the loss function of the model to guide the learning of model parameters.

4.4.3 Model Training Process

Through the above process, a physics-informed latent variable model, namely Spatio-Temporal Differential Equation Network (STDEN), is constructed. This model can be trained end-to-end via the backpropagation algorithm by minimizing the negative log-likelihood between the predicted traffic flow sequence and the observations in the training data. Compared with the training method of standard neural networks, the main difference of STDEN lies in that the forward propagation of its potential energy field differential equation network component needs to be computed with the help of the neural differential solver shown in Eq. (4.22). It is worth noting that this solver has been standardized and open-sourced, and can be easily integrated into existing computational frameworks.

For ease of understanding, a brief introduction to the solution of DE-Nets is provided here. First, the forward process typically involves solving ordinary differential equations (ODEs) using numerical methods, with the following two common explicit methods:

- Euler method (first-order accuracy):

$$\boldsymbol{h}(t+\Delta t) \approx \boldsymbol{h}(t) + \Delta t \cdot \mathcal{F}(\boldsymbol{h}(t), t, \theta). \tag{4.27}$$

- Fourth-order Runge-Kutta method (RK4, fourth-order accuracy):

$$\begin{aligned} \boldsymbol{h}(t+\Delta t) &\approx \boldsymbol{h}(t) + \frac{\Delta t}{6}(k_1 + 2k_2 + 2k_3 + k_4), \\ s.t.\ \ k_1 &= \mathcal{F}(\boldsymbol{h}(t), t, \theta), \\ k_2 &= \mathcal{F}\left(\boldsymbol{h}(t) + \frac{\Delta t}{2}k_1, t + \frac{\Delta t}{2}, \theta\right), \\ k_3 &= \mathcal{F}\left(\boldsymbol{h}(t) + \frac{\Delta t}{2}k_2, t + \frac{\Delta t}{2}, \theta\right), \\ k_4 &= \mathcal{F}\left(\boldsymbol{h}(t) + \Delta t k_3, t + \Delta t, \theta\right). \end{aligned} \tag{4.28}$$

During backpropagation, traditional automatic differentiation needs to cache all intermediate states, causing memory consumption to grow linearly with the "depth" (i.e., the number of integration steps). STDEN adopts the adjoint sensitivity method [4], which computes gradients in the reverse direction by solving the adjoint ODE. It only needs to store the initial state and final state, significantly reducing memory usage (with a space complexity of $O(1)$). The specific steps are as follows:

(i) Solve $\boldsymbol{h}(T)$ in the forward pass;
(ii) Construct the adjoint state $\boldsymbol{a}(t) = \frac{\partial \mathcal{L}}{\partial \boldsymbol{h}(t)}$, where $\mathcal{L}$ is the loss function;
(iii) Solve the adjoint equation in the reverse pass:

$$\frac{d\boldsymbol{a}(t)}{dt} = -\boldsymbol{a}(t)^{\top} \frac{\partial \mathcal{F}(\boldsymbol{h}(t), t, \theta)}{\partial \boldsymbol{h}(t)}. \tag{4.29}$$

(iv) Accumulate the parameter gradients:

$$\frac{\partial L}{\partial \theta} = -\int_{T}^{0} \boldsymbol{a}(t)^{\top} \frac{\partial \mathcal{F}(\boldsymbol{h}(t), t, \theta)}{\partial \theta} dt. \tag{4.30}$$

The above is the training process of the DE-Net. Through this process, the model can learn the spatio-temporal dynamic patterns in the data and achieve accurate prediction of future traffic flow states.

It is worth noting that in this model, the physical generation process of traffic flow is driven by a PEF and modeled by a PEF differential equation network based on deep residual GCN, thereby endowing the model with clear structural physical interpretability. In other words, unlike traditional deep learning models, this model is not a mere "black box" but can be regarded as a physics-guided generative model, thus promising to improve the prediction performance. In addition, the elaborately designed residual PEF differential equation network structure and the measurement mechanism for the uncertainty of the PEF may also further enhance the model's predictive power.

4.5 Demonstration and Outcome Analysis

This section demonstrates the empirical performance of the proposed methods and analyzes the resulting outcomes. It first outlines the experimental setup, followed by an overall comparison with baseline approaches. The section then investigates the impact of key components through ablation studies and concludes with an in-depth analysis of model behavior.

4.5.1 Evaluation Setup

This subsection sets up the evaluation by describing the dataset characteristics, the evaluation metrics, baseline methods for comparison, and key implementation details.

4.5.1.1 Data Description

In the experiments of this section, a real-world urban traffic dataset collected by the Beijing Municipal Commission of Transport is used to evaluate the performance of the model.

This dataset covers the trajectory data of 40,000 taxis in Beijing from April 1, 2015 to July 31, 2015 (a total of 4 months). These trajectories are mapped to the road network via the FMM[3] map-matching algorithm, and the number of taxis on each road segment is counted at 5 min time intervals to construct the traffic flow dataset. Ultimately, 288 time-step data points are obtained per day, providing a high spatio-temporal resolution basis for model training and evaluation.

Given the complexity of Beijing's road network, this study selects three sub-networks to construct experimental datasets, aiming to cover different urban functional areas without losing the diversity of the datasets. The selected sub-networks are as follows:

[3] Fast Map Matching algorithm: https://fmm-wiki.github.io/.

- GT-221 (Workers' Stadium, also known as Gongti): a famous entertainment district in Beijing, containing 221 road segments;
- WRS-393 (Surroundings of Beijing West Railway Station): a key transportation hub area in Beijing, containing 393 road segments;
- ZGC-564 (Zhongguancun, China's Silicon Valley): the largest business park in Beijing, containing 564 road segments.

The detailed information of these three datasets is provided in Table 4.2.

To verify the diversity of the datasets, statistical analyses were performed on the three datasets, as shown in Fig. 4.3. Specifically, we calculate the Pearson correlation coefficients between all pairs of data samples and plot their distribution. As can be seen from the results, there are significant differences in the correlation distributions among the three datasets, which indicates that they exhibit a certain degree of diversity in traffic flow patterns and is conducive to the comprehensive evaluation of the model's generalization ability.

In a road network $\mathcal{G} = (\mathcal{V}, \mathcal{E}, \boldsymbol{W})$ containing n intersections, the adjacency matrix $\boldsymbol{W} \in \mathbb{R}^{n \times n}$ is constructed by taking the length of the road $e_{ij} \in \mathcal{E}$ between intersections v_i and v_j as the weight. To enhance numerical stability and the generalization ability of the model, a Gaussian kernel function is used to normalize the weights, i.e.,

$$w_{ij} = \exp\left(-\frac{d_{ij}^2}{\sigma^2}\right), \tag{4.31}$$

Table 4.2 Statistical information of road network datasets for three districts in Beijing

Dataset	Time span	Time interval	# Samples	# Nodes	# Edges
GT-221	April 1, 2015 to July 31, 2015	5 min	34,560	203	221
WRS-393	April 1, 2015 to July 31, 2015	5 min	34,560	307	393
ZGC-564	April 1, 2015 to July 31, 2015	5 min	34,560	506	564

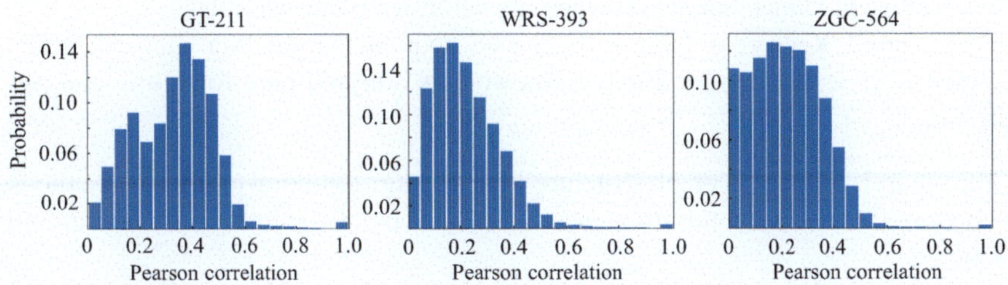

Fig. 4.3 Visualization of the correlation distribution of data samples

where w_{ij} denotes the normalized weight of the road e_{ij} between intersections v_i and v_j, d_{ij} is the actual length of the road e_{ij}, and σ is the standard deviation of road lengths. This normalization strategy can effectively smooth the weight distribution of the adjacency matrix, thereby improving the adaptability of the model to road networks of different scales.

In the experiments, each dataset is divided into a training set, a validation set, and a test set in a ratio of 7:1:2. For the multi-step prediction task, traffic flow data from the past hour (12 time steps) is employed as input to predict the traffic flow sequence for the next hour.

4.5.1.2 Evaluation Metrics and Baseline Models

The experiments adopt three commonly used prediction error evaluation metrics to assess the model performance, including: Mean Absolute Error (MAE), Root Mean Squared Error (RMSE), and Mean Absolute Percentage Error (MAPE). Among these metrics, MAE and RMSE measure from the perspective of absolute error, while MAPE measures from the perspective of relative error, which can more comprehensively reflect the prediction accuracy of the model. The smaller the values of all metrics, the better the prediction performance of the model.

To verify the effectiveness of STDEN, it is compared with 10 baseline models in the experiments, which mainly cover three categories.

The first category is time series modeling methods, where historical traffic flow sequences are treated as pure time series to predict future states without considering spatial information. They include:

- HA: It is the abbreviation of Historical Average. This method models historical traffic flow as a seasonal process and then uses the weighted average of data from the same time in the previous 4 weeks as the predicted value.
- VAR: Vector Autoregression (VAR) is a statistical method commonly used for multivariate time series modeling. It can capture the dynamic relationships of multiple variables over time, assuming that each variable depends not only on its own historical values but also on the past values of other variables.
- GRU: Gated Recurrent Unit is a classic sequence modeling method, and a sequence-to-sequence (Seq2Seq) framework is adopted here for multi-step prediction.

The second category is graph structure-based spatio-temporal prediction methods, which can effectively capture the spatial correlations of road networks and mainly include:

- DCRNN [14]: Diffusion Convolution Recurrent Neural Network is a graph convolution-based spatio-temporal prediction method. It captures spatial correlations through diffusion convolution layers and temporal dynamics through recurrent layers.

- STGCN [22]: Spatio-Temporal Graph Convolutional Network is a graph convolution-based spatio-temporal prediction method. It captures spatial correlations through graph convolution layers and temporal dynamics through temporal convolution layers.
- GWNET [21]: Graph WaveNet is a graph convolution-based spatio-temporal prediction method. It captures spatial correlations through graph convolution layers and temporal dynamics through temporal convolution layers.
- AGCRN [1]: Adaptive Graph Convolutional Recurrent Network is a graph convolution-based spatio-temporal prediction method. It captures spatial correlations through adaptive graph convolution layers and temporal dynamics through recurrent layers.
- MTGNN [20]: This model is a graph neural network framework designed for multivariate time series data. It combines graph convolution networks with mixed-hop propagation layers and dilated initial layers to capture spatial and temporal dependencies.

The third category is neural differential equation network-based methods, which can capture continuous-time dynamics and include:

- LODE [17]: Latent Ordinary Differential Equation (Latent-ODE) is a neural differential equation-based spatio-temporal prediction method. It generalizes recurrent neural networks to have continuous-time hidden dynamics defined by ordinary differential equations.
- OLSTM [11]: ODE-LSTM is a novel long short-term memory network with continuous-time output states, so its internal dynamic flow is modified into a continuous-time model.

For all deep learning baseline models, this chapter adopts the same model structure as the open-source code in the original papers, and selects the optimal hyperparameter settings based on the model performance on the validation set.

4.5.1.3 Model Implementation Details

In this chapter, the specific settings of the STDEN model are introduced in detail, mainly including the following two parts.

First is the configuration of the DE-Net part. The adaptive method $dopri5$ [6] is used to model the potential energy in the latent space, and a grid search is performed over the set of latent dimensions $\{1, 2, 4, 8\}$. Finally, 4 dimensions are selected based on the validation set results. The absolute tolerance and relative tolerance of the ODE solver are both set to 0.00001. In addition, DE-Net samples 3 initial states to generate 3 evolution trajectories, and obtains the average result at each evaluation time step by aggregating these trajectories. Meanwhile, DE-Net is modeled using a random walk ChebNet, with the filter size set to 64.

Next is the configuration of the encoder part. The GRU with the optimal hidden state selected on the validation set is used to encode the initial PEF distribution, and a fully connected layer containing 50 hidden units is adopted. STDEN is trained using the Adam optimizer with a learning rate annealing strategy, where the initial learning rate is set to 0.01 and decayed by a factor of 10 after every 20 training epochs. Meanwhile, gradient clipping technology is employed with the maximum norm set to 5. In addition, an early stopping strategy is applied during the training process, with the patience set to 20 and the maximum number of training epochs to 100. The batch size of all baseline models is set to 32.

4.5.2 Comparative Analysis of Overall Results

In the experiments, multiple repeated experiments were conducted for all deep learning models using 7 different random seeds, and the average results of all experiments were reported as the final results.

Tables 4.3, 4.4 and 4.5 present the comparison of future road network traffic flow prediction results for 15 min, 30 min, and 1 h obtained by different methods on three datasets. Due to the differences in the value ranges of various metrics, four significant figures are uniformly retained in the tables.

Based on the experimental results in all three tables, the following conclusions can be drawn. Firstly, spatio-temporal methods based on graph neural networks, such as AGCRN and GWNET, are generally superior to time series models like GRU. This indicates that modeling the spatial correlations of road networks is crucial in traffic prediction tasks. Compared with pure time series methods, spatio-temporal methods can better capture the complex dynamic characteristics of traffic flow.

Table 4.3 Comparison of road network traffic flow prediction results of different prediction methods on the GT-221 dataset

Task	15 min			30 min			1 h		
Metric	MAE	RMSE	MAPE	MAE	RMSE	MAPE	MAE	RMSE	MAPE
HA	1.324	1.835	61.47	1.324	1.835	61.47	1.324	1.835	61.47
VAR	1.125	1.581	52.34	1.141	1.620	52.03	1.164	1.835	51.51
GRU	0.988	1.511	42.77	1.000	1.541	42.95	1.018	1.578	43.29
STGCN	0.976	1.528	41.40	0.988	1.552	42.08	1.008	1.585	43.20
DCRNN	0.976	1.520	41.10	0.994	1.553	41.75	1.020	1.601	42.46
GWNET	0.967	1.516	40.70	0.980	1.540	41.23	1.006	1.581	42.12
AGCRN	0.968	1.502	41.18	0.981	1.525	41.73	1.004	1.561	42.79
MTGNN	0.962	1.505	40.88	0.976	1.531	41.40	1.003	1.576	42.36
LODE	0.981	1.531	42.11	0.992	1.544	42.71	1.011	1.588	43.28
OLSTM	0.978	1.522	42.87	0.986	1.561	41.80	1.007	1.579	43.22
STDEN	**0.865**	**1.317**	**36.72**	**0.872**	**1.328**	**37.37**	**0.896**	**1.360**	**39.04**

Table 4.4 Comparison of road network traffic flow prediction results of different prediction methods on the WRS-393 dataset

Task	15 min			30 min			1 h		
Metric	MAE	RMSE	MAPE	MAE	RMSE	MAPE	MAE	RMSE	MAPE
HA	1.239	1.735	64.08	1.239	1.735	64.08	1.239	1.735	64.08
VAR	1.070	1.509	55.51	1.097	1.568	55.72	1.139	1.653	56.01
GRU	0.823	1.393	35.52	0.837	1.429	36.00	0.856	1.477	36.10
STGCN	0.803	1.400	33.20	0.818	1.436	33.69	0.841	1.490	34.31
DCRNN	0.803	1.386	33.51	0.826	1.435	34.46	0.858	1.500	35.83
GWNET	0.801	1.390	33.50	0.818	1.427	33.95	0.845	1.488	34.37
AGCRN	0.799	1.382	33.28	0.814	1.414	33.96	0.841	1.468	34.93
MTGNN	0.792	1.376	32.91	0.807	1.411	33.34	0.834	1.469	34.01
LODE	0.818	1.402	35.33	0.827	1.433	35.77	0.853	1.482	35.87
OLSTM	0.809	1.391	35.28	0.818	1.428	35.12	0.848	1.478	35.14
STDEN	**0.730**	**1.241**	**31.62**	**0.737**	**1.244**	**32.04**	**0.745**	**1.266**	**33.25**

Table 4.5 Comparison of road network traffic flow prediction results of different prediction methods on the ZGC-564 dataset

Task	15 min			30 min			1 h		
Metric	MAE	RMSE	MAPE	MAE	RMSE	MAPE	MAE	RMSE	MAPE
HA	1.120	1.522	62.68	1.120	1.522	62.68	1.120	1.522	62.68
VAR	0.978	1.402	53.72	0.991	1.393	53.64	1.019	1.441	53.85
GRU	0.729	1.229	33.82	0.734	1.242	33.99	0.747	1.266	34.55
STGCN	0.714	1.218	32.63	0.724	1.241	33.13	0.738	1.272	33.78
DCRNN	0.715	1.230	31.99	0.727	1.258	32.34	0.746	1.298	32.86
GWNET	0.716	1.232	31.31	0.725	1.255	31.32	0.743	1.296	32.50
AGCRN	0.709	1.210	32.19	0.717	1.228	32.66	0.733	1.259	33.24
MTGNN	0.708	1.207	32.64	0.714	1.222	32.90	0.731	1.255	33.54
LODE	0.723	1.226	32.98	0.730	1.263	33.78	0.742	1.281	33.98
OLSTM	0.718	1.219	32.75	0.725	1.252	32.65	0.740	1.274	33.36
STDEN	**0.623**	**1.026**	**28.81**	**0.628**	**1.036**	**29.14**	**0.657**	**1.033**	**31.58**

Secondly, the STDEN model proposed in this chapter further improves the prediction accuracy on the basis of spatio-temporal methods. Compared with the suboptimal method, STDEN achieves average improvements of 10.29%, 13.49%, and 6.03% in the MAE, RMSE, and MAPE metrics, respectively. STDEN achieves the best performance across all evaluation metrics and prediction horizons. This indicates that the PEF differential equation designed under mechanism guidance has significant advantages in modeling continuous spatio-temporal dynamics.

Thirdly, network methods purely based on neural differential equations perform worse than spatio-temporal methods in traffic flow prediction. Although models based solely on neural differential equations have strong capabilities in modeling temporal dynamics, their overall performance is inferior to spatio-temporal methods because they ignore the crucial spatial correlations in traffic flow prediction. However, they still have certain advantages compared with traditional time series methods.

Finally, traditional methods (such as HA and VAR) have limited performance on complex spatio-temporal data because they cannot effectively capture the complex nonlinear characteristics of traffic flow, resulting in poor prediction performance. In addition, the prediction performance of the historical average method remains unchanged because its prediction results only depend on long-term historical patterns and are not affected by short-term data.

Taken together, the experimental results show that STDEN combines mechanism-guided differential equation modeling and the representation learning capability for spatio-temporal data, enabling it to more accurately capture the evolution laws of traffic flow and significantly outperform existing methods in traffic prediction tasks. This finding can inspire and encourage researchers in related fields to apply the idea of mechanism-guided model design to other spatio-temporal prediction tasks, thereby improving prediction accuracy and generalization performance.

4.5.3 Ablation Studies on Key Modules

This section mainly discusses the rationality of the model design and the sensitivity of the hyperparameters.

4.5.3.1 Module Ablation Analysis

To more intuitively verify the validity of the PEF differential equation in Eq. (4.14), this subsection conducts comparative experiments between STDEN and the following two variants, with the results presented in Fig. 4.4. The introductions of the two variants are as follows:

(i) *UnkP*: denotes the unknown physical equation, i.e., Unknown Physics. This variant replaces the PEF differential equation with a fully connected neural network to evaluate the performance change after removing physical modeling;
(ii) *IncP*: denotes the incomplete physical equation, i.e., Incomplete Physics. This variant ignores the energy volume factor $\boldsymbol{\phi}$ in the PEF differential equation, leading to incompleteness in physical modeling.

Figure 4.4 presents the experimental results of STDEN and its two variants regarding various metrics across all three datasets. The following phenomena can be observed from these results.

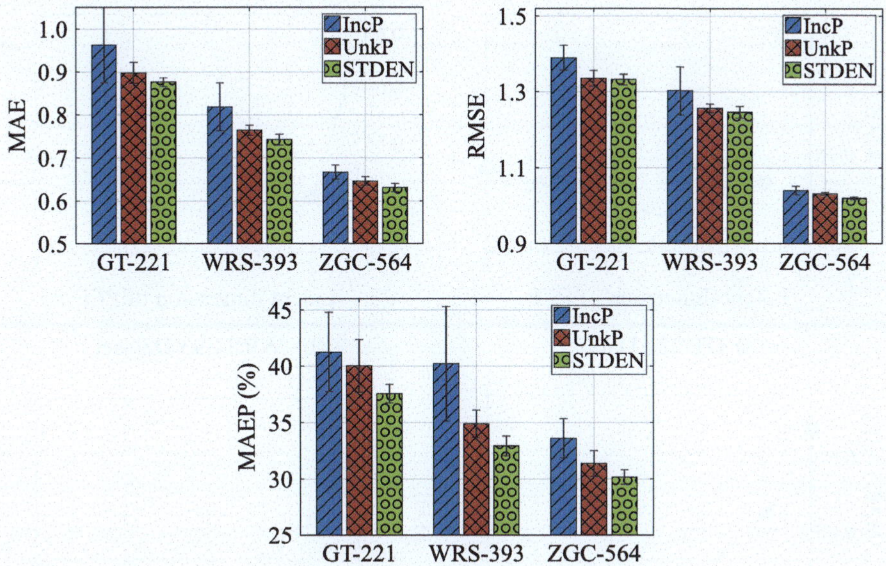

Fig. 4.4 Ablation experiment results of the STDEN model

First, STDEN significantly outperforms *UnkP* in all metrics, indicating that the introduction of the PEF into traffic flow modeling helps improve prediction accuracy.

Second, *IncP* performs better than *UnkP*, which further validates the effectiveness of the PEF differential equation in describing the spatio-temporal dynamics of traffic flow. Specifically, this equation characterizes the spatial diffusion process of energy, enabling *IncP* to capture the spatio-temporal variation trends of the PEF in road networks.

Finally, due to the lack of the energy volume factor $\boldsymbol{\phi}$, *IncP* achieves slightly lower performance than STDEN. This is because without $\boldsymbol{\phi}$, *IncP* can only rely on energy density to infer potential energy, which is similar to attempting to calculate the mass of an object solely by its density when the volume is unknown, thereby increasing the uncertainty in potential energy estimation.

4.5.3.2 Hyperparameter Impact Analysis

To analyze the impact of hyperparameters on model performance, this section of experiments investigates the key components of STDEN, including the DE-Net for the PEF and the encoder GRU.

Specifically, the latent dimension of the PEF is searched over the set $\{1, 2, 4, 8\}$, while the number of hidden units in the GRU is searched over the set $\{16, 32, 64, 128\}$, so as to evaluate the influence of different hyperparameter settings on model perfor-

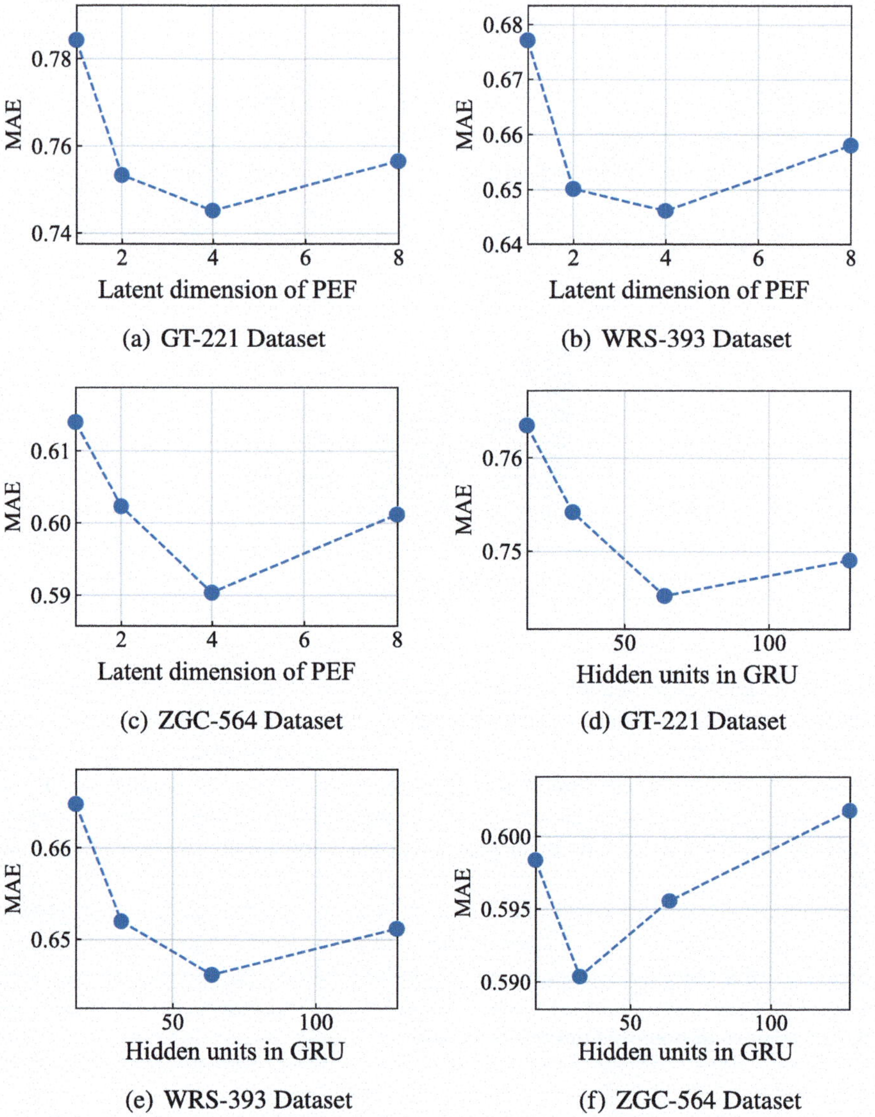

Fig. 4.5 Experimental results of hyperparameters for the STDEN model

mance. Note that the performance evaluation metric here is the MAE on the validation set.

Figure 4.5 presents the experimental results of hyperparameter sensitivity on all datasets. From the results, it can be observed that appropriately increasing the latent dimension of the PEF can significantly improve the prediction accuracy on the validation set. However, when the latent dimension is excessively large, redundant information may be introduced, making the model more prone to memorizing the training data and causing overfitting, which leads to a decline in generalization ability.

Similarly, when adjusting the number of hidden units in the GRU, the same trend is observed: a moderate increase in the number of hidden units can enhance the prediction accuracy, but exceeding a certain threshold will impair the generalization ability of the model.

Based on the results of the above parameter sensitivity analysis, the STDEN model proposed in this chapter finally adopts the hyperparameters with the best performance on the validation set as its model configuration to conduct the experiments in Tables 4.3, 4.4, and 4.5, as well as subsequent experiments for in-depth analysis of model behavior, thereby ensuring the optimal balance between model complexity and generalization performance.

4.5.4 In-Depth Analysis of Model Behavior

This subsection provides a deeper examination of model behavior through qualitative and quantitative analyses. It includes visualizations of prediction results, an interpretability analysis to reveal underlying mechanisms, and an assessment of computational cost in relation to prediction accuracy, offering insights into model efficiency and practical applicability.

4.5.4.1 Visualization of Prediction Results

To further analyze the prediction results, two road segments are randomly selected in this part of the experiment, and their prediction results in the 1-hour-ahead traffic flow prediction task are presented. To compare the performance of the potential energy differential equation, the prediction results of UnkP—the unknown physics variant of STDEN that replaces the potential energy equation with a neural network—are also incorporated herein.

Figure 4.6 presents the visualization results, where Ground Truth represents the true values of future traffic flow. Two key findings can be derived from the results. First, under the guidance of the potential energy equation, STDEN can more effectively identify the increasing and decreasing patterns of traffic flow. For instance, in box a1 of Fig. 4.6a, the true traffic flow first rises and then falls, and STDEN successfully captures this change, whereas UnkP yields an opposite trend prediction. Similarly, in box a2, STDEN identifies the start and end times of the traffic flow peak, while UnkP fails to accurately characterize them. This advantage mainly originates from the ability of the potential energy equation to model spatial dependencies, allowing STDEN to leverage the potential energy changes of neighboring nodes to generate more reasonable predictions.

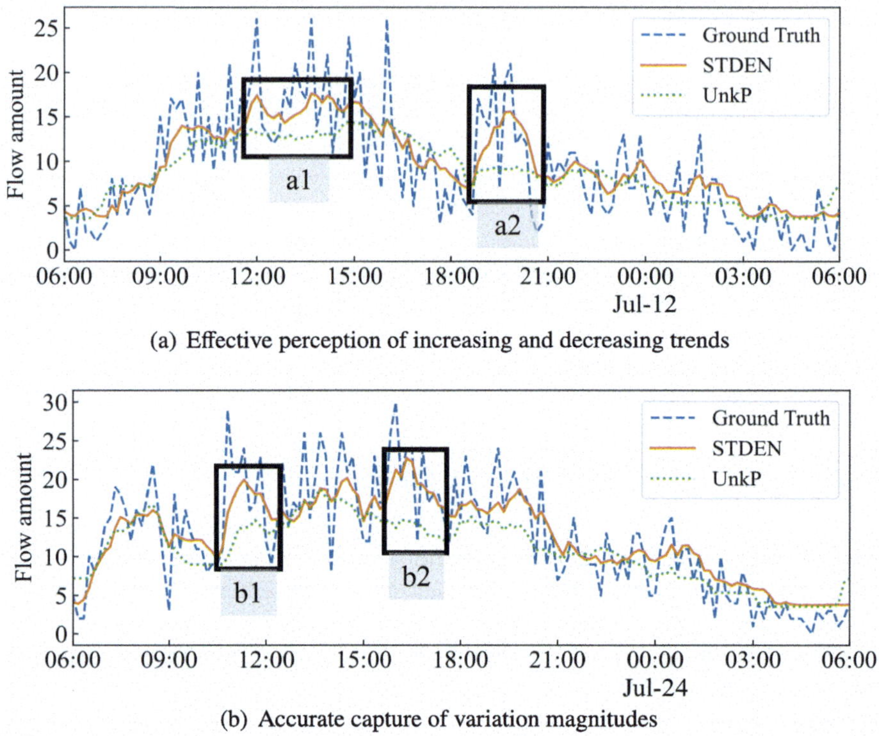

Fig. 4.6 Visualization of traffic flow prediction results

On the other hand, the differential equation network STDEN, which is designed under the guidance of mechanistic models, can more accurately characterize the magnitude of traffic flow changes (Fig. 4.6b). For example, in box b1, the magnitude of change predicted by STDEN is close to 80%, while UnkP can only capture about 40% of the change. This indicates that STDEN exhibits stronger robustness when dealing with traffic flow fluctuations, and can more accurately measure the magnitude of flow changes, thereby improving prediction accuracy and practical applicability.

4.5.4.2 Interpretability Analysis

The in-depth analysis experiments on model behavior in this section first conduct a qualitative analysis of the superiority of STDEN in traffic flow prediction tasks, and then further explore the reasons for its excellent performance.

To this end, taking the GT-221 dataset as a case study, the potential energy distribution learned by the model and the real traffic flow are visualized in Fig. 4.7, where the heatmap represents the distribution of the PEF: the redder the color, the larger the potential energy value, and the bluer the color, the smaller the potential energy value; on the other hand, the arrows represent traffic flow, and the magnitude of the traffic flow is reflected by the color and length of the arrows. In addition, a variant model of STDEN, *UnkP*, is selected as a comparison in the figure; this variant model relies solely on neural network learning and does not include the constraints of the PEF differential equation.

Subfigures (a) and (d) of Fig. 4.7 respectively show the potential energy distributions learned by STDEN and UnkP, with the real traffic flow at the corresponding time overlaid. To further analyze how potential energy drives traffic flow, a detailed investigation was conducted on the complex traffic loop area in Subfigures (b) and (c) of Fig. 4.7. The results demonstrate that traffic flow always moves from high-potential energy regions to low-potential energy regions along the road network, and the potential energy gradient can well reflect the relative magnitude of traffic flow. This phenomenon indicates that under the constraint of the differential equation of the PEF, STDEN has successfully learned the physical mechanism driving traffic flow, endowing the model's prediction results with interpretability.

In contrast, in the absence of guidance from the differential equation of the PEF, the "energy" learned by UnkP lacks a clear physical meaning and cannot reasonably describe the traffic flow mechanism. For example, in Fig. 4.7e and f, some traffic flows exhibit the phenomenon of flowing from low-potential-energy regions to high-potential-energy regions, which violates the physical constraints in the energy transfer process, indicating that UnkP, which relies solely on neural network learning, cannot accurately capture the intrinsic laws of traffic flow.

In summary, the experimental results indicate that STDEN, under the guidance of the PEF differential equation, can effectively model the physical mechanisms of traffic flow (Fig. 4.7) and generate more accurate prediction results in experiments (see the experimental results in Tables 4.3, 4.4 and 4.5). This advantage can be attributed

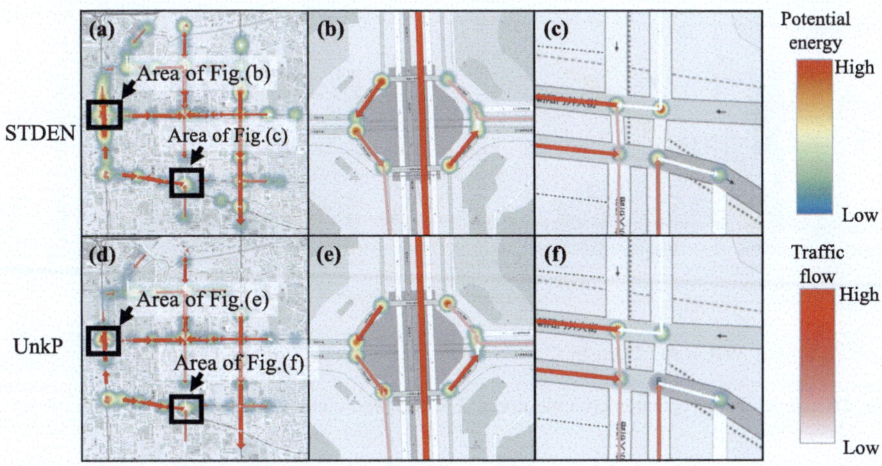

Fig. 4.7 Visualization of the learned PEF and actual traffic flow on the GT-221 dataset

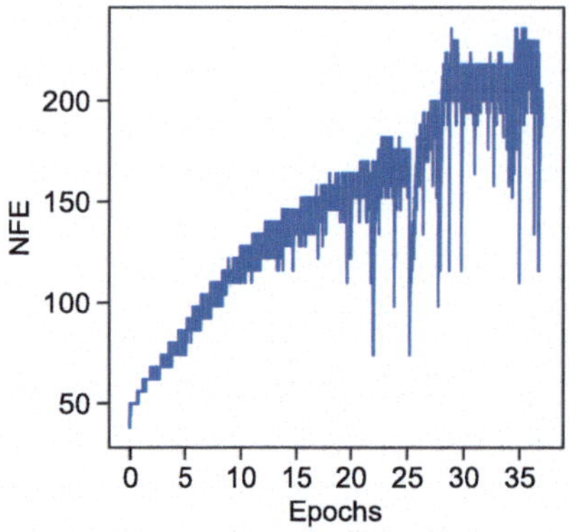

(a) GT-221 Dataset

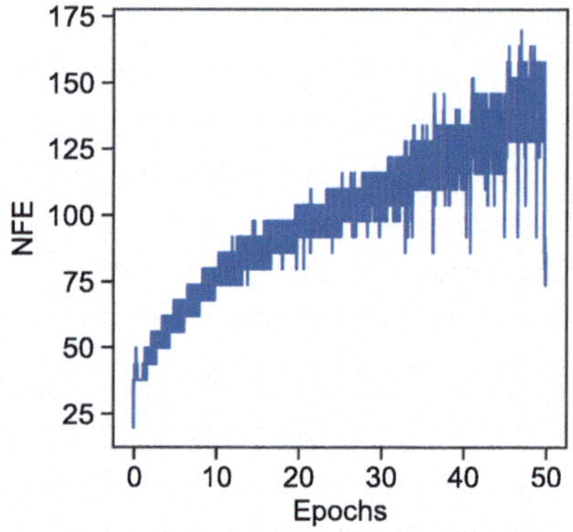

(b) WRS-393 Dataset

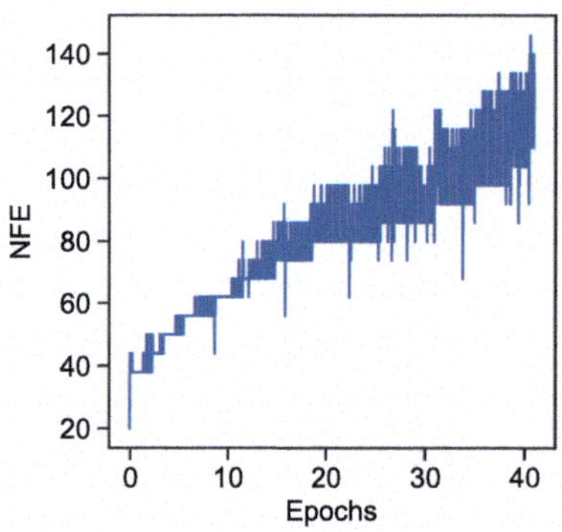

(c) ZGC-564 Dataset

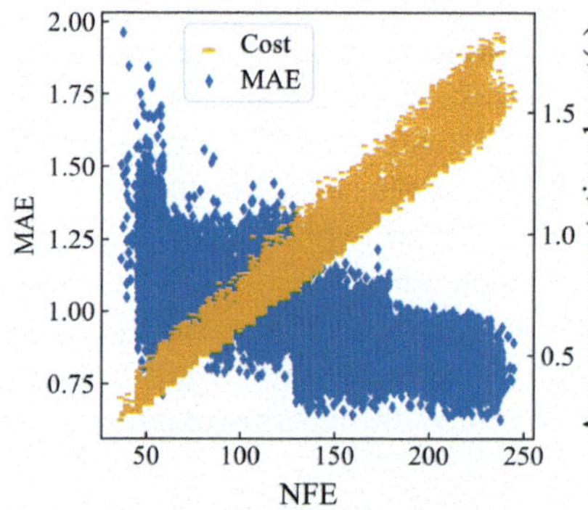

(d) GT-221 Dataset

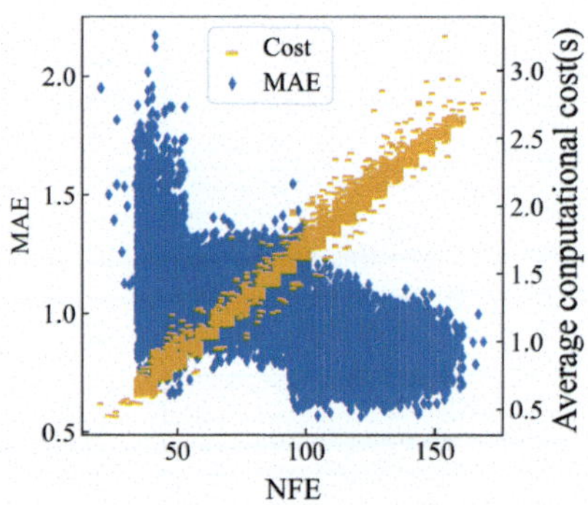

(e) WRS-393 Dataset

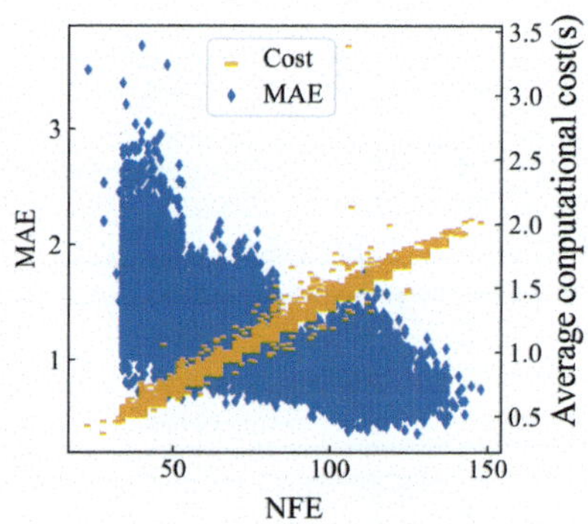

(f) ZGC-564 Dataset

Fig. 4.8 Relationship between computational cost and prediction accuracy of the STDEN model to the fact that STDEN combines the complementary strengths of physics-based methods and data-driven methods, thereby enhancing the physical consistency and prediction accuracy of the model.

4.5.4.3 Computational Cost and Prediction Accuracy

In the solution process of differential equation networks, this chapter employs an ODE solver, enabling the model to dynamically balance prediction accuracy and computational cost.

As shown in Fig. 4.8a, b, and c, the Number of Function Evaluation (NFE) continuously increases with the increase in the number of training epochs and does not exhibit a stable phase, which is mainly due to the early stopping strategy set during the training process.

It should be noted that the variation of NFE is affected by task complexity. For example, the traffic flow prediction task on the GT-221 dataset is more challenging than other tasks because the traffic flow in this dataset fluctuates greatly. Therefore, its NFE value is approximately 80% higher than that of ZGC-564.

In addition, from the results of Fig. 4.8d, e and f, it can be observed that a higher NFE usually corresponds to a lower prediction error. This characteristic allows the model to adjust the computational cost according to requirements during the inference phase to respond more quickly to emergencies, thus having important practical value in real-world applications.

4.6 Summary

This chapter proposes a Spatio-Temporal Differential Equation Network (STDEN) model based on the continuity equation, which aims to address the matching problem between traditional discrete modeling methods and the requirements of continuous traffic flow prediction. First, by introducing a potential energy field (PEF) with continuous-time characteristics as the underlying force driving traffic flow in the road network, and based on the continuity equation of energy, a differential equation is derived to describe the physical mechanism of the continuous-time road network PEF and simulate the dynamic transmission process of traffic flow in the road network.

Next, to solve the problem that traditional discrete hidden-layer neural networks cannot express dynamic hidden states, a novel spatio-temporal differential equation network with continuous hidden layers is designed based on the PEF differential equation. By converting the initial value problem of the differential equation into a continuous propagation process of network layers, this network enhances the model's ability to model the evolution of continuous-time traffic flow. Meanwhile, it integrates deep learning into physical mechanism modeling to ensure that the traffic flow prediction process satisfies the mechanistic constraints of the PEF.

Extensive experiments conducted on three real-world traffic datasets demonstrate the effectiveness of the proposed STDEN, especially compared with spatio-temporal prediction models that do not model temporal continuity. Case studies further verify that the STDEN model can capture the mechanisms of urban traffic and generate accurate predictions with physical significance.

The STDEN model is suitable for tasks that require accurate capture of the continuous changes in traffic flow over the temporal dimension. Meanwhile, the traffic flow data in this model is defined on the road network, which differs from grid-based regional traffic flow. Due to the involvement of continuous-time modeling and differential equation solving, this model may incur high computational overhead; especially in large-scale urban traffic networks or high-frequency data prediction tasks, the demand for computational resources is relatively high.

References

1. Bai, L., Yao, L., Li, C., Wang, X., Wang, C.: Adaptive graph convolutional recurrent network for traffic forecasting. In: Proceedings of the 34th International Conference on Neural Information Processing Systems (NeurIPS'20), pp. 17804–17815 (2020)
2. Batchelor, G.K.: An introduction to fluid dynamics. Cambridge University Press (2000)
3. Bronstein, M.M., Bruna, J., LeCun, Y., Szlam, A., Vandergheynst, P.: Geometric deep learning: Going beyond Euclidean data. IEEE Signal Processing Magazine **34**(4), 18–42 (2017)
4. Chen, T.Q., Rubanova, Y., Bettencourt, J., Duvenaud, D.: Neural ordinary differential equations. In: Proceedings of the 32nd International Conference on Neural Information Processing Systems (NeurIPS'18), pp. 6572–6583 (2018)
5. Cho, K., van Merriënboer, B., Gulcehre, C., Bahdanau, D., Bougares, F., Schwenk, H., Bengio, Y.: Learning phrase representations using RNN encoder–decoder for statistical machine translation. In: Proceedings of the 2014 Conference on Empirical Methods in Natural Language Processing, pp. 1724–1734 (2014)
6. Dormand, J.R., Prince, P.J.: A family of embedded Runge-Kutta formulae. Journal of Computational and Applied Mathematics **6**(1), 19–26 (1980)
7. He, K., Zhang, X., Ren, S., Sun, J.: Deep residual learning for image recognition. In: Proceedings of the 2016 IEEE Conference on Computer Vision and Pattern Recognition (CVPR'16), pp. 770–778 (2016)
8. Jackson, J.D.: Classical electrodynamics. John Wiley and Sons (1998)
9. Ji, J., Wang, J., Jiang, Z., Jiang, J., Zhang, H.: STDEN: Towards physics-guided neural networks for traffic flow prediction. In: Proceedings of the 36th AAAI Conference on Artificial Intelligence (AAAI'22), pp. 4048–4056 (2022)
10. Kingma, D.P., Welling, M.: Auto-encoding variational Bayes. In: Proceedings of the 2nd International Conference on Learning Representations (ICLR'14), pp. 1–14 (2014)
11. Lechner, M., Hasani, R.M.: Learning long-term dependencies in irregularly-sampled time series. In: Proceedings of the 36th International Conference on Neural Information Processing Systems (NeurIPS'22), pp. 1–19 (2022)
12. Li, T., Zhang, J., Bao, K., Liang, Y., Li, Y., Zheng, Y.: AutoST: Efficient neural architecture search for spatio-temporal prediction. In: Proceedings of the 26th ACM SIGKDD Conference on Knowledge Discovery and Data Mining (KDD'20), pp. 794–802 (2020)
13. Li, W., Wang, X., Zhang, Y., Wu, Q.: Traffic flow prediction over muti-sensor data correlation with graph convolution network. Neurocomputing **427**(1), 50–63 (2021)

14. Li, Y., Yu, R., Shahabi, C., Liu, Y.: Diffusion convolutional recurrent neural network: Data-driven traffic forecasting. In: Proceedings of the 6th International Conference on Learning Representations (ICLR'18), pp. 1–16 (2018)
15. Lienhard, J.I., Lienhard, J.: DOE fundamentals handbook: Thermodynamics, heat transfer, and fluid flow. Washington, DC: US Department of Energy (2008)
16. Lu, Y., Zhong, A., Li, Q., Dong, B.: Beyond finite layer neural networks: Bridging deep architectures and numerical differential equations. In: Proceedings of the 35th International Conference on Machine Learning (ICML'18), pp. 3282–3291 (2018)
17. Rubanova, Y., Chen, R.T., Duvenaud, D.: Latent ODEs for irregularly-sampled time series. In: Proceedings of the 33rd International Conference on Neural Information Processing Systems (NeurIPS'19), pp. 5320–5330 (2019)
18. Ruthotto, L., Haber, E.: Deep neural networks motivated by partial differential equations. Journal of Mathematical Imaging and Vision **62**(3), 352–364 (2019)
19. Wu, N., Zhao, W.X., Wang, J., Pan, D.: Learning effective road network representation with hierarchical graph neural networks. In: Proceedings of the 26th ACM SIGKDD Conference on Knowledge Discovery and Data Mining (KDD'20), pp. 6–14 (2020)
20. Wu, Z., Pan, S., Long, G., Jiang, J., Chang, X., Zhang, C.: Connecting the dots: Multivariate time series forecasting with graph neural networks. In: Proceedings of the 26th ACM SIGKDD Conference on Knowledge Discovery and Data Mining (KDD'20), pp. 753–763 (2020)
21. Wu, Z., Pan, S., Long, G., Jiang, J., Zhang, C.: Graph WaveNet for deep spatial-temporal graph modeling. In: Proceedings of the 28th International Joint Conference on Artificial Intelligence (IJCAI'19), pp. 1907–1913 (2019)
22. Yu, B., Yin, H., Zhu, Z.: Spatio-temporal graph convolutional networks: A deep learning framework for traffic forecasting. In: Proceedings of the 27th International Joint Conference on Artificial Intelligence (IJCAI'18), pp. 3634–3640 (2018)

Chapter 5
Basis Vector Representation Model

Abstract In urban traffic scenarios, spatio-temporal data distribution shift modeling is crucial for improving the adaptability and reliability of traffic prediction models under different spatio-temporal conditions. Because traffic conditions are affected by factors such as holidays and weather, their spatio-temporal distribution changes significantly, causing traditional correlation learning models to deteriorate when the training data do not match the actual application scenario. Therefore, this chapter introduces a spatio-temporal basis vector representation model based on causal modeling. This model represents any unknown confounding factors in the form of basis vectors, thereby accurately modeling their impact on future data distribution and improving the model's robustness and generalization ability under distribution shift scenarios.

5.1 Modeling Motivation

In the domain of urban traffic, the distribution of spatio-temporal data is pervasively influenced by a myriad of factors. These encompass the secular evolution of transportation infrastructure, dynamic shifts in human mobility patterns, stochastic interference from abrupt events, and the impact of external environmental factors such as weather and public holidays. These elements collectively induce dynamic variations in the spatio-temporal distribution of traffic states—a phenomenon termed spatio-temporal data distribution shift.

When traffic prediction models fail to effectively account for this distribution shift, the patterns learned during the training phase are often rendered obsolete in real-world applications. This discrepancy precipitates a severe degradation in predictive accuracy and fundamentally compromises the decision-making quality of intelligent transportation systems.

Consequently, modeling spatio-temporal data distribution shift is an essential undertaking for enhancing the robustness and generalization capabilities of traffic prediction. Specifically, it enables models to adapt to long-term trend changes, such as the impact of newly constructed metro lines or road network optimizations on traffic states. Concurrently, it enhances the model's resilience to abrupt events, including

J. Wang and J. Ji, *Mechanism-Driven Explainable Urban Spatio-Temporal Prediction*, Big Data Management, https://doi.org/10.1007/978-981-92-0662-9_5

the impacts of adverse weather, traffic accidents, or large-scale public gatherings. Moreover, given that urban traffic data exhibits significant spatial heterogeneity—wherein traffic patterns vary markedly across regions—accounting for distribution shift is paramount for maintaining stable predictive performance under diverse scenarios, thereby offering more precise support for intelligent traffic management and optimization.

Existing traffic forecasting models typically learn a mapping from historical states X to future states Y [2, 14, 25]. Despite their success, this dominant paradigm has three limitations for real-world deployment. First, it implicitly assumes a stationary $X \rightarrow Y$ relation, so models trained on past data may fail when deployment conditions shift. Second, recent approaches that traverse confounder values [8, 24] treat confounders as discrete and predefined, whereas traffic is an open-world setting with continuous and partly unobserved influences, making enumeration of such factors intractable. Third, while most confounder-based modeling focuses on learning invariant relations, practical systems also require scenario-conditional adaptation to handle complex spatiotemporal variability.

To address distribution shifts induced by confounders C, this chapter proposes the Spatio-TEmporal basis Vector Embedding model (STEVE), grounded in causal modeling principles [15]. It tackles the challenge of unknown and continuous confounders by representing arbitrary confounders using basis vectors. It leverages self-supervised learning (SSL) to ensure the learned representations have sufficient capacity to model these confounders, thereby enabling the disentanglement of confounding influences and the precise modeling of their impact on future data distributions during prediction.

Firstly, advancing beyond the prevalent correlation-based paradigm, this chapter re-analyzes and reframes the forecasting problem from a causal modeling perspective. Most conventional works follow the same paradigm: modeling the direct causal relationship $X \rightarrow Y$ (Fig. 5.1a). This assumes a stable, directed causal link between historical states X and future states Y, which data-driven approaches can model effectively. However, this assumption does not universally apply. Spatiotemporal dependence between X and Y is intrinsically affected by external factors such as rain, accidents, or holidays. In causal theory, these are confounders C that simultaneously affect X and Y, causing a shift in the original $X \rightarrow Y$ relationship (Fig. 5.1b). This fundamental issue limits model generalization during extreme weather or emergencies, impairing urban resilience. For example, heavy snow (a confounder) may

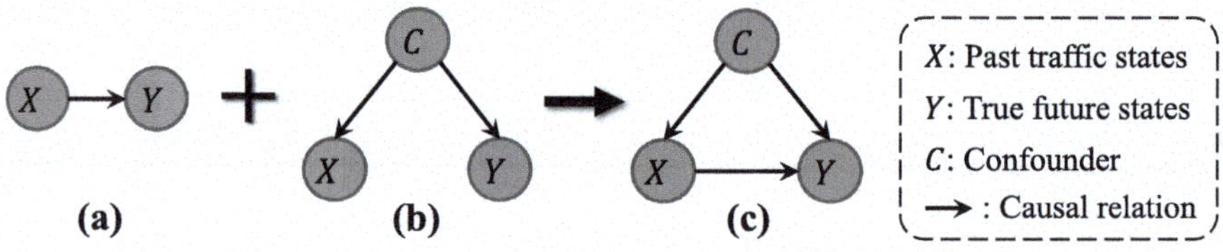

Fig. 5.1 The spatiotemporal traffic forecasting problem from a causal perspective

induce more cautious driving, leading to severe congestion during off-peak hours and altering the underlying dependency between X and Y. Ignoring this dynamically changing relationship compromises the model's performance in snowy-day traffic management. To this end, this chapter incorporates confounders into the traffic forecasting framework, redefining the task to account for these external factors.

Secondly, to characterize continuous or unknown novel confounding factors and associate them with the spatio-temporal samples from the current input, we propose a confounding factor extraction algorithm based on the concept of basis vectors and SSL methods to learn representations of arbitrary confounding factors. Specifically, by adopting the basis vector concept, a confounder is represented as a linear combination of basis confounding vectors. This transformation simplifies the challenge of direct confounder extraction into learning these fundamental bases. We train a neural network to learn a set of basis vectors, establishing a *basis confounder bank*, rather than focusing on specific individual confounders. This bank allows any arbitrary confounder—continuous or discrete, predefined or not—to be represented by a linear combination of its bases, with weights dynamically generated via a cross-attention mechanism between the input sample and the bank.

To ensure the basis confounder bank's expressive power for diverse confounders, we implement SSL with three auxiliary tasks: spatial position classification, temporal index recognition, and traffic load prediction. These tasks integrate the spatial, temporal, and semantic information relevant to the confounders, enabling the model to learn basis vectors with robust representation capabilities.

Finally, to capture confounder-invariant spatiotemporal relationships while simultaneously modeling confounder influence, this chapter designs a confounder-invariant relationship disentanglement module based on adversarial learning. This module is essential for constructing a robust traffic prediction model that captures both crucial relationship types.

Both confounder-related and confounder-invariant relationships are vital, but since explicit labels for invariant relationships are infeasible, we leverage the mutually exclusive nature of the two relationship types and employ adversarial learning. A Gradient Reversal Layer (GRL) is introduced to adversarially optimize the SSL loss. This procedure forces the learned invariant representations to become indistinguishable across different confounders, ensuring the successful learning of confounder-invariant relationships at the semantic level. Furthermore, a second mechanism based on mutual-information minimization loss achieves separation between the confounder and invariant representations at the distribution level, ensuring the model captures the statistical independence required for robust disentanglement.

Based on this complete decoupling, the confounder representations (from the basis confounder bank) and the confounder-invariant representations are converted into their respective traffic-state predictions. A dedicated fusion module then intelligently combines these two predictive outcomes to generate the final result, significantly enhancing the generalization ability and robustness of the traffic prediction model.

5.2 Basic Concepts and Problem Definition

This section formalizes urban traffic as a spatio-temporal process on traffic graphs and states the canonical forecasting problem. The task is subsequently reframed via Structural Causal Models (SCM) [18] to accommodate confounding, thereby making explicit the causal structure, codifying identification conditions, and guiding estimator design.

Definition 5.1 (*Traffic Graph*) Let a set of traffic entities (e.g., spatial regions, road segments) be denoted as $\mathcal{V} = \{v_n \mid 1 \leq n \leq N\}$. The traffic graph is defined as $\mathcal{G} = (\mathcal{V}, \boldsymbol{A})$, where $\boldsymbol{A} \in \mathbb{R}^{N \times N}$ is the adjacency matrix. If there is an edge connecting node v_m and v_n, then $a_{m,n} = 1$; otherwise, $a_{m,n} = 0$.

Definition 5.2 (*Traffic State*) Given a traffic entity v_n, assume it has F traffic state features (e.g., average speed, traffic inflow/outflow). The traffic state of v_n at time t is $\boldsymbol{x}_{n,t} \in \mathbb{R}^F$. The traffic states of all entities at time t form $\boldsymbol{X}_t \in \mathbb{R}^{N \times F}$. Historical states from $t - T + 1$ to t are $\mathcal{X}_t = (\boldsymbol{X}_{t-T+1}, \ldots, \boldsymbol{X}_t) \in \mathbb{R}^{T \times N \times F}$, and the future state is $\boldsymbol{Y}_{t+1} = \boldsymbol{X}_{t+1}$.

Definition 5.3 (*Classical Traffic Forecasting Problem*) Given $\mathcal{X}_t$, the goal is to predict $\boldsymbol{Y}_{t+1}$ via

$$\hat{\boldsymbol{Y}}_{t+1} = f(\mathcal{X}_t; \boldsymbol{\Theta}), \tag{5.1}$$

where $\hat{\boldsymbol{Y}}_{t+1}$ is the prediction of $\boldsymbol{Y}_{t+1}$, $f(\cdot)$ is the prediction function, and $\boldsymbol{\Theta}$ are learnable parameters.

To elucidate the mechanisms underlying temporal variation in traffic data, the Structural Causal Model (SCM) [18] is adopted. Let X denote the historical state and Y the future state. Their association can arise from (i) a direct causal effect $X \to Y$ and (ii) a confounder C that influences both X and Y, i.e., $X \leftarrow C \to Y$ (see Fig. 5.1). Most prior work models only case (i), yielding the confounder-irrelevant conditional $\Pr^{(i)}(Y \mid X)$. A central limitation is that Eq. (5.1) leaves the influence of C implicit, which can impair generalization.

If confounders are taken into account, the observational conditional is

$$\Pr^{(c)}(Y|X) = \Pr(Y|X, C)\Pr(C|X). \tag{5.2}$$

The confounder-related forecasting can be divided into two steps. First, learn confounder representations from $\mathcal{X}_t$ to estimate $\Pr(C \mid X)$:

$$\boldsymbol{C}_t = g^{(c)}(\mathcal{X}_t), \tag{5.3}$$

where $g^{(c)}(\cdot)$ is the confounder extractor. Second, predict the future state with the extracted confounders:

$$\hat{\boldsymbol{Y}}_{t+1}^{(c)} = f^{(c)}(\mathcal{X}_t, \boldsymbol{C}_t; \boldsymbol{\Theta}^{(c)}), \tag{5.4}$$

where $f^{(c)}(\cdot)$ is the confounder-relevant predictor with parameters $\boldsymbol{\Theta}^{(c)}$.

Definition 5.4 (*Traffic Forecasting Problem Considering Confounders*) By combining the relationships $\Pr^{(c)}(Y|X)$ and $\Pr^{(i)}(Y|X)$, the overall probability relationship is obtained as

$$\Pr(Y|X) = \Pr^{(c)}(Y|X) + \Pr^{(i)}(Y|X). \tag{5.5}$$

Accordingly, the spatio-temporal forecasting problem is reformulated as

$$\hat{\boldsymbol{Y}}_{t+1} \;=\; f^{(c)}\big(\boldsymbol{\mathcal{X}}_t, \boldsymbol{C}_t; \boldsymbol{\Theta}^{(c)}\big) \;+\; f^{(i)}\big(\boldsymbol{\mathcal{X}}_t; \boldsymbol{\Theta}^{(i)}\big), \tag{5.6}$$

where $f^{(c)}(\cdot)$ and $f^{(i)}(\cdot)$ are the confounder-relevant and -irrelevant predictors with parameters $\boldsymbol{\Theta}^{(c)}$ and $\boldsymbol{\Theta}^{(i)}$, respectively.

5.3 Basis Vector Model for Spatio-Temporal Confounders

The goal is to implement Eq. (5.6) via the STEVE model, shown in Fig. 5.2. The model takes historical observations as input, predicts future traffic, and uses self-supervised signals to promote confounder representation learning. In Fig. 5.2, TCL denotes Temporal Convolutional Layer, GCL denotes Graph Convolutional Layer, and COSSL denotes Confounder-Oriented Self-Supervised Learning (see Fig. 5.4 in Sect. 5.3.2). The Confounder Extractor architecture is in Fig. 5.3 (Sect. 5.3.1). For brevity, sample indices are omitted.

Next, we detail the overall pipeline and its principal modules: a confounder-representation generator, a confounder-aware self-supervised learning module, and a disentanglement module for confounder-irrelevant relationships.

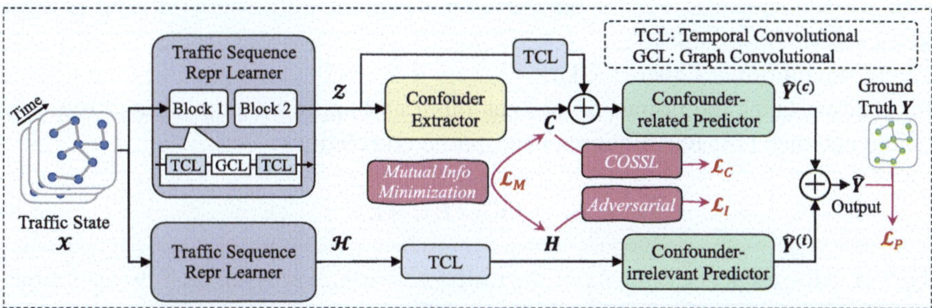

Fig. 5.2 Overview of the STEVE model architecture

5.3.1 Confounder Representation Generation

This component aims to generate confounder representations from the input traffic data. To achieve this goal, we first employ a traffic sequence representation learner to embed dynamic temporal dependencies and time-varying spatial relations into latent representations. Next, we introduce a learnable confounder extractor that adaptively distills complex and dynamic confounding factors from the latent space. Finally, the resulting confounder representations are calibrated by a confounder-oriented self-supervised learning module (see Sect. 5.3.2) to ensure that they faithfully capture the confounders of interest.

5.3.1.1 Traffic Sequence Representation Learner

The traffic sequence representation learning (TSRL) module transforms the input traffic sequence $\mathcal{X}_t \in \mathbb{R}^{T\times N\times F}$ into a latent representation $\mathcal{Z}_t \in \mathbb{R}^{T\times N\times D}$. TSRL leverages a temporal convolutional layer and a graph convolutional layer to capture temporal patterns and spatial dependencies across spatio-temporal locations, which are described below.

Temporal Convolutional Layer (TCL)

Given the traffic state sequence $\mathcal{X}_t = (\boldsymbol{X}_{t-T+1}, \ldots, \boldsymbol{X}_t) \in \mathbb{R}^{T\times N\times F}$, TCL applies a 1D convolution along the temporal dimension[23] to extract time-aware traffic representations:

$$(\boldsymbol{E}_{t-T_1+1}, \ldots, \boldsymbol{E}_t) = \mathrm{TCL}(\boldsymbol{X}_{t-T+1}, \ldots, \boldsymbol{X}_t), \tag{5.7}$$

where $\boldsymbol{E}_t \in \mathbb{R}^{N\times D}$ denotes the representation matrix at time step t, T_1 is the length of the output sequence, N is the number of nodes in the traffic network, and D is the representation dimension.

Graph Convolutional Layer (GCL)

This module takes the output of TCL as input and applies a graph-based message passing network [14] to further capture spatial correlations:

$$\boldsymbol{S}_t = \mathrm{GCL}(\boldsymbol{E}_t, \boldsymbol{A}), \tag{5.8}$$

where $\boldsymbol{A}$ is the adjacency matrix of the traffic network. Applying GCL to each time-aware representation $\boldsymbol{E}_t$ yields enhanced traffic representations $(\boldsymbol{S}_{t-T_1+1}, \ldots, \boldsymbol{S}_t)$, thereby strengthening spatio-temporal feature expressiveness.

To jointly model temporal and spatial dependencies, we follow the design in [25] and construct TSRL as a two-block architecture. Each block consists of a TCL → GCL → TCL stack, as illustrated in Fig. 5.2. The final output of TSRL is given by

$$\mathcal{Z}_t = (\boldsymbol{Z}_{t-T+1}, \ldots, \boldsymbol{Z}_t) = \mathrm{TSRL}(\mathcal{X}_t, \boldsymbol{A}). \tag{5.9}$$

In the subsequent confounder extractor, $\mathcal{Z}_t \in \mathbb{R}^{T \times N \times D}$ will be used to derive confounder representations.

Furthermore, to model relations that are independent of confounders, we employ another TSRL model with a disjoint set of parameters to produce an additional representation $\mathcal{H}_t \in \mathbb{R}^{T \times N \times D}$. This representation is used for correlation modeling in Sect. 5.3.3.

5.3.1.2 Confounder Extractor

This section implements the function $g^{(c)}(\cdot)$ in Eq. (5.3), which extracts confounder representations from historical traffic states. We introduce the motivation and the method as follows.

Motivation and Idea

As discussed earlier, $g^{(c)}(\cdot)$ aims to approximate the conditional distribution $\Pr(C \mid X)$ in Eq. (5.2), where C and X denote the random variables of confounders and historical traffic states, respectively. However, directly approximating $\Pr(C \mid X)$ is challenging for two main reasons: (i) C may follow a complex mixture distribution conditioned on multiple latent factors, making the modeling difficult; (ii) the support of C can be unbounded (or effectively infinite), further increasing approximation complexity.

To address these challenges, inspired by prior work [19], we introduce a set of basis variables to represent $\Pr(C \mid X)$:

$$\Pr(C \mid X) = \sum_{k=1}^{K} \alpha^{(k)} \phi\left(\beta^{(k)} \mid X\right), \tag{5.10}$$

where K is the number of basis variables, and $\alpha^{(k)}$ denotes the membership (mixture) weight of C belonging to the k-th basis component, satisfying $\sum_{k=1}^{K} \alpha^{(k)} = 1$. Here, $\phi(\beta^{(k)} \mid X)$ is the conditional probability function of the k-th basis variable given X. It can be viewed as a basic confounding factor that composes more complex confounders, and the mixture weights $\boldsymbol{\alpha} = (\alpha^{(1)}, \ldots, \alpha^{(K)})$ are learnable.

Since $\boldsymbol{\alpha}$ varies continuously, it can in principle characterize infinitely many confounders. For example, if the basic confounders include rush hours, rainfall, and holidays, then by adjusting $\boldsymbol{\alpha}$ one can represent compound confounders such as "rush hours on rainy weekdays".

Method Details

Following the above intuition, we construct an adaptive basis-confounder bank $\boldsymbol{B} = (\boldsymbol{b}^{(1)}, \ldots, \boldsymbol{b}^{(K)})^\top \in \mathbb{R}^{K \times D}$, as shown in Fig. 5.3. This bank represents different basic confounders, where D is the embedding dimension and each basis vector $\boldsymbol{b}^{(k)}$ instantiates $\phi(\beta^{(k)} \mid X)$.

Given the traffic latent representation sequence $(\mathcal{Z}_1, \ldots, \mathcal{Z}_t)$ in chronological order, we map it to $(\tilde{\boldsymbol{B}}_1, \ldots, \tilde{\boldsymbol{B}}_t)$ via

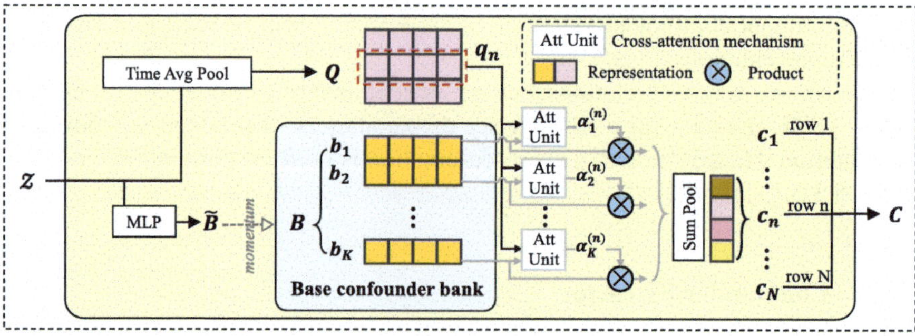

Fig. 5.3 Overview of the confounder extractor

$$\tilde{\boldsymbol{B}}_t = \text{MLP}\big(\text{Flatten}_2(\mathcal{Z}_t)\big). \tag{5.11}$$

Here, $\text{Flatten}_2(\cdot)$ flattens the first two dimensions of $\mathcal{Z}_t \in \mathbb{R}^{T \times N \times D}$, yielding a matrix of shape $(TN) \times D$. We then apply a multi-layer perceptron MLP($\cdot$) to compress the first dimension and obtain $\tilde{\boldsymbol{B}}_t \in \mathbb{R}^{K \times D}$, i.e., an adaptive basis-confounder bank at time step t.

We next use $\tilde{\boldsymbol{B}}_t$ to update the confounder bank $\boldsymbol{B}$. A straightforward strategy is an online update, i.e., $\boldsymbol{B}_t = \tilde{\boldsymbol{B}}_t$, which preserves the environmental context of the current sample. However, it discards historical information and is thus inconsistent with our goal of modeling confounders with both short- and long-term regularities. Ideally, $\boldsymbol{B}_t$ should sense the current environment while integrating information from previous samples to better capture long-term patterns.

To this end, we adopt a momentum update:

$$\boldsymbol{B}_t = \gamma \boldsymbol{B}_{t-1} + (1-\gamma)\tilde{\boldsymbol{B}}_t, \tag{5.12}$$

where γ is the momentum coefficient controlling the retention of historical information. A larger γ results in smoother updates, whereas a smaller γ makes the bank more responsive to the current sample.

Moreover, to improve the effectiveness of the confounder bank, we initialize basis vectors randomly and apply PCA whitening [1] to remove correlations among basis vectors, thereby enhancing their expressive capacity for diverse confounders.

Given the basis-confounder bank, we compute the mixture weights $\alpha^{(k)}$ in Eq. (5.10) via cross-attention [21] between the current sample and the bank, enabling adaptive modeling of dynamic confounding factors. Specifically, for the n-th traffic entity v_n, we first construct a confounder query vector $\boldsymbol{q}_{n,t} \in \mathbb{R}^D$:

$$\boldsymbol{q}_{n,t} = \frac{1}{T} \sum_{\tau=t-T+1}^{t} z_{n,\tau}, \tag{5.13}$$

where $z_{n,\tau}$ is the latent representation of entity v_n at time τ. By averaging latent representations over the historical window, $\boldsymbol{q}_{n,t}$ aggregates temporal context and yields a stable query for downstream modeling.

We then compute attention weights $\alpha_{n,t}^{(k)}$ between $\boldsymbol{q}_{n,t}$ and the basis vector $\boldsymbol{b}_t^{(k)}$:

$$\alpha_{n,t}^{(k)} = \frac{\exp\left(u\left(\boldsymbol{q}_{n,t}, \boldsymbol{b}_t^{(k)}\right)\right)}{\sum_{j=1}^{K} \exp\left(u\left(\boldsymbol{q}_{n,t}, \boldsymbol{b}_t^{(j)}\right)\right)}, \tag{5.14}$$

where $u : \mathbb{R}^D \times \mathbb{R}^D \to \mathbb{R}$ is a scoring function that measures the relevance between the query vector $\boldsymbol{q}_{n,t}$ and each basis vector $\boldsymbol{b}_t^{(k)}$. We parameterize $u(\cdot, \cdot)$ using an MLP to enhance nonlinear modeling capacity, thereby more accurately characterizing the confounder distribution of each traffic entity.

Finally, we derive the confounder representation as a weighted combination of basis vectors:

$$\boldsymbol{c}_{n,t} = \sum_{k=1}^{K} \alpha_{n,t}^{(k)} \boldsymbol{b}_t^{(k)}, \tag{5.15}$$

where $\boldsymbol{c}_{n,t}$ is the n-th row of the confounder representation matrix $\boldsymbol{C}_t \in \mathbb{R}^{N \times D}$, corresponding to traffic entity v_n at time step t. This matrix is generated from the input sample $\mathcal{X}_t$ through its latent representation $\mathcal{Z}_t$. When a new sample $\mathcal{X}_{t+1}$ arrives, we feed $\mathcal{Z}_{t+1}$ into Eq. (5.11) and repeat the above procedure to obtain $\boldsymbol{c}_{n,t+1}$, enabling continual updates of confounder representations and maintaining adaptability to evolving environments.

Remark After training, the basis-confounder bank $\boldsymbol{B}$ forms a vector space in which each confounder can be regarded as a point uniquely determined by its mixture weights $(\alpha^{(1)}, \ldots, \alpha^{(K)})$. Since this space contains infinitely many points, the bank can represent any confounder within the learned space—whether discrete or continuous, predefined or previously unseen. Therefore, the proposed mechanism not only adapts to known confounders but also generalizes to latent, unannotated, and complex confounding factors present in the data. Moreover, during inference, the bank can be adjusted according to the latent environment of incoming samples, making it more aligned with the current observations. This further improves generalization under unseen confounders and enhances robustness in non-stationary traffic environments.

5.3.2 Confounder-Oriented Self-supervised Learning

The confounder-oriented self-supervised learning (COSSL) module optimizes the confounder representation $\boldsymbol{C}$ using self-supervised signals that are informative of confounders, as illustrated in Fig. 5.4.

Since it is difficult to enumerate all potential confounders, we propose to use a set of representative confounders as self-supervised signals to strengthen the ability of

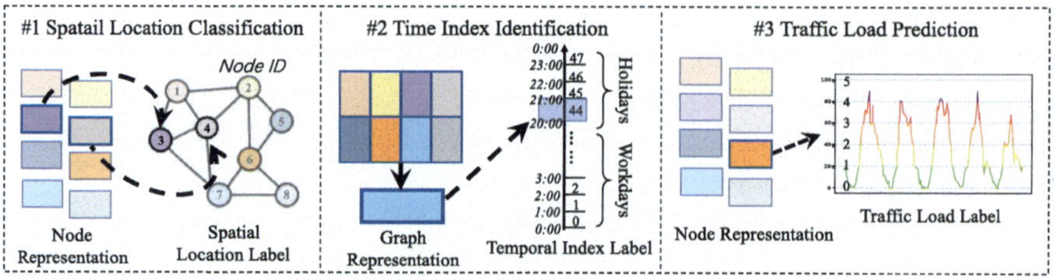

Fig. 5.4 The confounder-oriented self-supervised learning (COSSL) module

$\boldsymbol{C}$ to encode confounder-related information[1]. Leveraging the unique properties of traffic data, we summarize latent factors that influence traffic conditions into three conceptual categories: *time, space, and semantics*. We then select representative and easily accessible variables from each category, including spatial location, temporal index, and traffic capacity. These confounders serve as learning signals for the following three self-supervised tasks.

5.3.2.1 Spatial Location Classification

The spatial location of a traffic entity reflects its environmental context and can act as a confounder that influences the temporal dependency $(\boldsymbol{x}_{t-T+1}, \ldots, \boldsymbol{x}_t) \rightarrow \boldsymbol{x}_{t+1}$. Different locations may induce different dependency patterns; for example, transportation hubs can exhibit markedly distinct temporal dynamics compared with office districts. We therefore design a spatial location classification task to capture location-specific environmental information.

Specifically, for a traffic entity $v_n \in \mathcal{V}$, we assign a unique location label based on its node ID and represent it using a one-hot vector $\boldsymbol{y}_n^{(1)} \in \{0, 1\}^N$, where $y_{n,m}^{(1)} = 1$ if $m = n$ and $y_{n,m}^{(1)} = 0$ otherwise. The task is optimized using a cross-entropy loss:

$$\ell_{sl}(C) = \frac{1}{N} \sum_{n=1}^{N} \ell_{sl}^{(n)} = -\frac{1}{N} \sum_{n=1}^{N} \sum_{m=1}^{N} y_{n,m}^{(1)} \log \left(\hat{y}_{n,m}^{(1)} \right), \tag{5.16}$$

where $\hat{y}_{n,m}^{(1)}$ is the predicted probability that the n-th traffic entity belongs to class m. This probability is obtained from the m-th entry of $\hat{\boldsymbol{y}}_n^{(1)} = g_1(\boldsymbol{c}_n) \in \mathbb{R}^N$. The function $g_1(\cdot)$ is implemented as a two-layer MLP followed by a Softmax normalization, and $\boldsymbol{c}_n$ denotes the n-th row of the confounder representation matrix $\boldsymbol{C}$.

[1] For notational simplicity, we omit the sample index t in $\boldsymbol{C}_t$.

5.3.2.2 Temporal Index Identification

Time-varying confounders such as weather conditions and holidays can substantially affect the distribution of traffic data. For example, on holidays, peak-hour fluctuations may become less pronounced than on weekdays, leading to distinct traffic patterns. To better exploit such temporal factors, we design a temporal index identification task to enhance the model's awareness of time-dependent dynamics.

Concretely, we partition one day into 24 time slots and define separate categories for weekdays and holidays, resulting in a total of $I_t = 48$ temporal indices. Given a traffic sample $(\boldsymbol{\mathcal{X}}, \boldsymbol{Y})$, we use the temporal index of $\boldsymbol{Y}$ as the ground-truth label and represent it as a one-hot vector $\boldsymbol{y}^{(2)} \in \{0, 1\}^{I_t}$. The objective of temporal index identification is

$$\ell_{ti}(C) = -\sum_{i=1}^{I_t} y_i^{(2)} \log\left(\sigma(\hat{y}^{(2)})_i\right), \tag{5.17}$$

where $\sigma(\cdot)$ is the Softmax function and $\hat{\boldsymbol{y}}^{(2)} = \frac{1}{N}\sum_{n=1}^{N} g_2(\boldsymbol{c}_n)$ is the predicted temporal-index vector. The function $g_2(\cdot)$ is a two-layer MLP that strengthens the temporal information encoded in $\boldsymbol{c}_n$.

5.3.2.3 Traffic Load Prediction

Traffic load reflects semantic congestion status of traffic entities and can serve as a confounder that strongly influences future traffic evolution. For instance, when traffic load approaches saturation, congestion is more likely to occur, leading to speed drops and reduced inflow/outflow in subsequent time steps. We therefore design a traffic load prediction task to incorporate dynamic load information into the confounder representation.

Specifically, we approximate the load capacity of the n-th traffic entity using its historical maximum traffic flow:

$$CP_n = \max(\{\boldsymbol{x}_{t,n}\}_1^{\tau}) \in \mathbb{R}^F, \tag{5.18}$$

where τ denotes the total number of time steps in the training set and $\max(\cdot)$ extracts the maximum value over the sequence. We then discretize flow into six load levels and compute the flow load of entity n as

$$\boldsymbol{y}_n^{(3)} = \lceil 5\, \boldsymbol{y}_n / CP_n \rceil \in \{0, \ldots, 5\}^F, \tag{5.19}$$

where $\boldsymbol{y}_n$ is the label in the main forecasting task. Given the imbalance of real-world traffic load, we use mean squared error (MSE) as the objective:

$$\ell_{tl}(\boldsymbol{C}) = \frac{1}{N}\sum_{n=1}^{N} \left\| g_3(\boldsymbol{c}_n) - \boldsymbol{y}_n^{(3)} \right\|_2^2, \tag{5.20}$$

where $g_3(\cdot)$ is a two-layer MLP load-prediction head and $\boldsymbol{c}_n$ is the confounder representation.

Note that although traffic flow load is discretized into levels, it naturally preserves an ordinal structure. Compared with a classification loss, MSE is more suitable here because it explicitly captures numerical distances between levels, enabling a more faithful modeling of load dynamics.

Finally, we jointly optimize the above three self-supervised losses to train $\boldsymbol{C}$ so that it integrates diverse confounder information. The overall objective of COSSL is

$$\mathcal{L}_C = \sum_{u \in \{sl, ti, tl\}} \ell_u(\boldsymbol{C}). \tag{5.21}$$

Remark The representative confounders selected from the temporal, spatial, and semantic perspectives provide effective self-supervised signals that guide STEVE to capture richer confounder-related information. Training with these auxiliary tasks can improve generalization by encouraging more robust representations, allowing the model to remain adaptive even under unseen confounding factors.

5.3.3 Confounder-Irrelevant Relationship Disentanglement

As in Eq. (5.6), accurately characterizing the confounder-irrelevant relationship is as crucial as capturing confounder-relevant dynamics. Since the confounder-irrelevant representation should minimize leakage of confounder information, we separate it from $\boldsymbol{C}$ via semantic and distributional disentanglement.

Recall the TSRL module in Sect. 5.3.1, used here to model the confounder-irrelevant relationship and generate $\mathcal{H}_t \in \mathbb{R}^{T \times N \times D}$. Applying the TCL operation (Eq. (5.7)) along time yields $\boldsymbol{H}_t \in \mathbb{R}^{N \times D}$.

We then disentangle $\boldsymbol{H}$ from $\boldsymbol{C}$ at semantic and distributional levels, making $\boldsymbol{H}$ confounder-agnostic. For simplicity, we omit sample indices and write $\boldsymbol{H}_t \to \boldsymbol{H}$, $\boldsymbol{C}_t \to \boldsymbol{C}$.

5.3.3.1 Adversarial Disentanglement

To effectively remove confounding information from $\boldsymbol{H}$, we adopt adversarial disentanglement (see Fig. 5.5). The generator g_θ comprises TSRL and TCL and outputs $\boldsymbol{H}$; the discriminator g_ψ performs the same confounder-related self-supervised tasks as in Sect. 5.3.2.

Unlike learning $\boldsymbol{C}$, a Gradient Reversal Layer (GRL) [12] is inserted between generator and discriminator when learning $\boldsymbol{H}$. In the forward pass, GRL leaves the

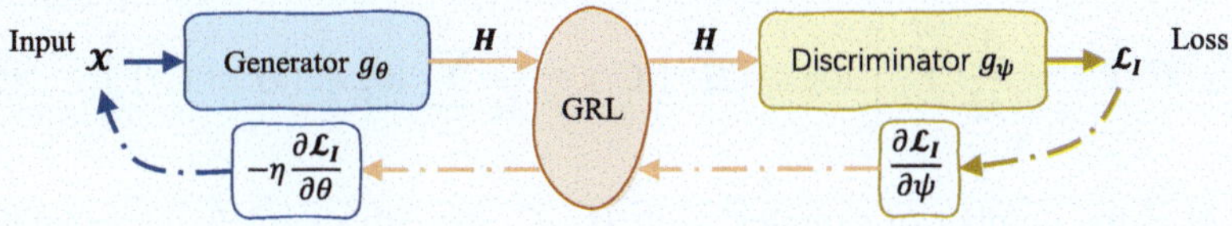

Fig. 5.5 Adversarial learning-based disentanglement with a Gradient Reversal Layer (GRL)

input unchanged:

$$\boldsymbol{H} = \mathrm{GRL}_\eta(\boldsymbol{H}). \tag{5.22}$$

In backpropagation, the gradient to the generator is multiplied by $-\eta$:

$$\frac{\partial\, \mathrm{GRL}_\eta}{\partial\, \boldsymbol{H}} = -\eta \boldsymbol{I}, \tag{5.23}$$

with $\boldsymbol{I}$ the identity. Thus the generator is pushed away from confounder-discriminative features, promoting disentanglement so $\boldsymbol{H}$ remains semantically consistent across confounding environments.

The objective is

$$\mathcal{L}_I(\psi, \theta) = \sum_{u \in \mathcal{U}} \ell_u(\mathrm{GRL}_\eta(\boldsymbol{H})) = \sum_{u \in \mathcal{U}} \ell_u\big(\mathrm{GRL}_\eta(g_\theta(\boldsymbol{\mathcal{X}}))\big), \tag{5.24}$$

where $\mathcal{U} = \{sl, ti, tl\}$ denotes the self-supervised tasks in Sect. 5.3.2, and ψ are discriminator parameters. Training follows

$$\psi = \arg\min_{\psi} \ \mathcal{L}_I(\psi, \theta), \quad \theta = \arg\max_{\theta} \ \mathcal{L}_I(\psi, \theta). \tag{5.25}$$

Thus ψ minimizes and θ maximizes $\mathcal{L}_I$, establishing an adversarial game.

5.3.3.2 Mutual Information Minimization

Independence constraints are commonly used in disentanglement [3]. We therefore minimize mutual information (MI) between $\boldsymbol{H}$ and $\boldsymbol{C}$:

$$\mathcal{L}_M = \mathrm{MI}(\boldsymbol{H}, \boldsymbol{C}) = \mathbb{E}_{\Pr(\boldsymbol{H}, \boldsymbol{C})}\left[\log \frac{\Pr(\boldsymbol{H}, \boldsymbol{C})}{\Pr(\boldsymbol{H})\Pr(\boldsymbol{C})}\right]. \tag{5.26}$$

Because the marginals and the joint are unknown, MI is not directly computable. We adopt an estimation method using the CLUB upper bound [5]:

$$\widehat{\mathrm{MI}}_{\mathrm{ub}}(\boldsymbol{H}, \boldsymbol{C}) = \frac{1}{M}\sum_{i=1}^{M}\left[\log q_{\theta}(\boldsymbol{H}_i \mid \boldsymbol{C}_i) - \frac{1}{M}\sum_{j=1}^{M}\log q_{\theta}\big(\boldsymbol{H}_j \mid \boldsymbol{C}_i\big)\right], \tag{5.27}$$

where M is the number of samples, with $\boldsymbol{H}_i, \boldsymbol{C}_i$ the i-th sample's representations. The conditional $q_{\theta}(\boldsymbol{H} \mid \boldsymbol{C})$ is a variational approximation to $\Pr(\boldsymbol{H} \mid \boldsymbol{C})$, modeled as $\mathcal{N}(\mu_{\boldsymbol{C}}, \sigma^2_{\boldsymbol{C}})$. An MLP predicts $\{\mu_{\boldsymbol{C}}, \sigma^2_{\boldsymbol{C}}\}$:

$$\{\mu_{\boldsymbol{C}}, \sigma^2_{\boldsymbol{C}}\} = \mathrm{MLP}(\boldsymbol{C}; \boldsymbol{\Theta}_{\mathrm{mlp}}), \tag{5.28}$$

with learnable $\boldsymbol{\Theta}_{\mathrm{mlp}}$.

Finally, we use $\widehat{\mathrm{MI}}_{\mathrm{ub}}(\boldsymbol{H}, \boldsymbol{C})$ in Eq. (5.27) as a surrogate for $\mathrm{MI}(\boldsymbol{H}, \boldsymbol{C})$ in Eq. (5.26) to implement MI minimization. This enforces an independence constraint between $\boldsymbol{H}$ and $\boldsymbol{C}$, and experiments (see analysis section) show it enhances disentanglement.

5.3.4 *Model Training Process*

This section first forecasts the confounder-relevant and confounder-irrelevant components separately. Their predictions are then fused to compute the traffic-state forecasting loss. Finally, this loss is aggregated with the previously defined objectives to jointly optimize the model, updating parameters so as to preserve predictive accuracy while promoting de-confounding.

5.3.4.1 Prediction and Fusion

On the one hand, based on $\boldsymbol{C}_t$ and $\mathcal{Z}_t$, the confounder-aware prediction function $f_c(\cdot)$ in (5.6) can be constructed, and confounder-aware traffic state prediction is performed using the following formula:

$$\hat{\boldsymbol{Y}}^{(c)}_{t+1} = \mathrm{MLP}\big(\boldsymbol{C}_t + \mathrm{TCL}(\mathcal{Z}_t)\big), \tag{5.29}$$

where the definition of TCL is provided in (5.7), and its role is to integrate information from $\mathcal{Z}_t$ along the temporal dimension. The MLP consists of a two-layer fully connected network and is used for non-linear mapping.

On the other hand, based on the confounder-irrelevant representation $\boldsymbol{H}_t$, the confounder-irrelevant prediction function $f_i(\cdot)$ in (5.6) can be constructed, defined as:

$$\hat{\boldsymbol{Y}}^{(i)}_{t+1} = \mathrm{MLP}(\boldsymbol{H}_t). \tag{5.30}$$

The prediction module operates solely on the deconfounded representation $\boldsymbol{H}_t$, thereby ensuring that outputs are not influenced by confounding factors. In practice, the magnitude of confounder effects varies across regions—for example, peak-hour patterns strongly affect office and residential districts but have relatively minor impact on parks and entertainment areas. Moreover, even within a single region, different traffic states (e.g., inflow and outflow) can exhibit distinct sensitivities to the same confounders. Motivated by these observations, this chapter proposes a heterogeneity-aware fusion method that dynamically adjusts the influence of confounders in the prediction task, with the following form:

$$\hat{\boldsymbol{Y}}_{t+1} = \boldsymbol{\Lambda}_1 \odot \hat{\boldsymbol{Y}}^{(c)}_{t+1} + \hat{\boldsymbol{Y}}^{(i)}_{t+1}, \tag{5.31}$$

where $\hat{\boldsymbol{Y}}_{t+1}$ is the final traffic state prediction result, and $\odot$ denotes the element-wise Hadamard product. The weight matrix $\boldsymbol{\Lambda}_1$ is responsible for dynamically adjusting the influence intensity of the confounders, defined as:

$$\boldsymbol{\Lambda}_1 = \text{Sigmoid}(\boldsymbol{C}_t \boldsymbol{W}_c), \tag{5.32}$$

where $\boldsymbol{W}_c \in \mathbb{R}^{D \times F}$ is a learnable parameter matrix. Through $\boldsymbol{\Lambda}_1$, the model can adaptively adjust the contribution of confounders to the final prediction based on different regions and traffic states, thereby enhancing the flexibility and accuracy of the prediction.

5.3.4.2 Training Objective

Based on the final prediction result $\hat{\boldsymbol{Y}}_{t+1}$, the error of the traffic state prediction task can be calculated as the loss for the prediction task:

$$\mathcal{L}_P = \frac{1}{NF} \sum_{i=1}^{N} \sum_{j=1}^{F} \left| y_{i,j} - \hat{y}_{i,j} \right|, \tag{5.33}$$

where $\hat{y}_{i,j}$ is an element in $\hat{\boldsymbol{Y}}_{t+1} \in \mathbb{R}^{N \times F}$, $y_{i,j}$ represents the corresponding ground-truth value, N denotes the number of traffic entities, and F is the number of predicted traffic states.

Finally, to achieve joint optimization of confounder disentanglement and traffic state prediction, $\mathcal{L}_C$ (confounder representation loss), $\mathcal{L}_M$ (mutual information minimization loss), $\mathcal{L}_I$ (adversarial disentanglement loss), and $\mathcal{L}_P$ (traffic prediction error) are integrated into a unified optimization objective, yielding the overall loss function:

$$\mathcal{L}_O = \mathcal{L}_P + \gamma_1 \mathcal{L}_C + \gamma_2 \mathcal{L}_M + \gamma_3 \mathcal{L}_I, \tag{5.34}$$

where $\gamma_1, \gamma_2, \gamma_3$ are hyperparameters used to balance the impact of different loss terms on the model optimization. Through this joint optimization objective, the model can enhance predictive performance while achieving de-confounding, realizing robust traffic state modeling under dynamic distributions.

5.4 Implementation and Observations

This section reports implementation details and key empirical observations. It first describes the evaluation setup, followed by an overall comparison and analysis of results. The section then presents ablation studies on critical modules and concludes with an in-depth analysis of model behavior to highlight practical insights.

5.4.1 Evaluation Setup

This subsection describes the experimental setup used for empirical evaluation, including the characteristics of the dataset, evaluation metrics, benchmark methods used for comparison, and key implementation details.

5.4.1.1 Data Description

To evaluate the effectiveness of the proposed method, this chapter conducts experiments on four real-world traffic datasets, including NYCBike1, NYCBike2, NYC-Taxi, and BJTaxi [14].

These datasets consist of trajectory data from millions of taxis or bicycles, covering thousands of time steps and hundreds of geographical regions. Their statistical information is detailed in Table 5.1. Among them, NYCBike1 and NYCBike2 are bicycle datasets, while NYCTaxi and BJTaxi are taxi datasets.

The bicycle datasets record the rental demand in different regions, while the taxi datasets contain the number of inflow and outflow vehicles within specific time intervals. The specific descriptions of each dataset are as follows:

- NYCTaxi: Records 30-minute-level taxi flow data in New York City from January 1, 2015, to March 1, 2015.
- NYCBike1: Covers hourly bicycle rental data from April 1, 2014, to September 30, 2014.

Table 5.1 Statistical information of bicycle and taxi datasets from New York and Beijing

Dataset	NYCBike1	NYCBike2	NYCTaxi	BJTaxi
Data type	Bike rentals		Taxi GPS data	
Time Interval	1 hour	30 min	30 min	30 min
Num of Regions	16×8	10×20	10×20	32×32
Num of Taxis/Bikes	6.8k+	2.6m+	22m+	34k+
Num of Samples	4,392	2,880	2,880	5,596

- NYCBike2: Contains 30-minute-level bicycle rental data in New York City from July 1, 2016, to August 29, 2016.
- BJTaxi: Collects 30-minute-level taxi data in Beijing from March 1, 2015, to June 30, 2015.

These datasets span multiple cities, temporal granularities, and transportation modes, enabling a rigorous assessment of the model's adaptability across diverse traffic environments.

In all datasets, the traffic graph is constructed from inter-regional adjacency to encode spatial dependencies among areas. For the prediction task at time t , two categories of historical data are used as input to fully exploit temporal structure in the traffic state series. Specifically, the input comprises:

- *Short-term historical information*: Data from the 4 hours preceding time period t, used to capture recent dynamic trends.
- *Periodic historical information*: Data from 2 hours before and after the time periods $t - T_{\text{day}}$, $t - 2T_{\text{day}}$, and $t - 3T_{\text{day}}$, where T_{day} is the number of time steps in a day. This input aims to model the periodic features of daily traffic patterns.

To construct the training samples, a sliding window strategy is used to segment the data, and the dataset is divided into training, validation, and test sets in a 7:1:2 ratio, ensuring the model can be effectively evaluated on different data splits.

5.4.1.2 Evaluation Metrics and Baseline Models

The experiments employ Mean Absolute Error (MAE) and Mean Absolute Percentage Error (MAPE) as evaluation metrics [8, 25]. These metrics are widely used in spatio-temporal forecasting, and lower values indicate better predictive performance.

In light of the documented limitations of traditional statistical models and shallow machine learning methods for spatio-temporal data modeling [2, 13], the STEVE model is compared against 13 state-of-the-art baselines. These baselines primarily fall into the following three categories.

- GNN-based spatio-temporal prediction methods:
  - **STGCN** [25]: A graph convolution-based model that combines 1D convolutions to capture spatial and temporal correlations.
  - **AGCRN** [2]: Enhances classic graph convolution with an adaptive adjacency matrix and embeds it into an RNN structure.
  - **ASTGNN** [13]: Incorporates self-attention to model dynamics and heterogeneity in both temporal and spatial dimensions.
  - **HimNet** [9]: Captures spatio-temporal heterogeneity via learned spatial and temporal reps, with a meta-parameter pool for specialization.
- Disentanglement-based spatio-temporal prediction methods:

- COST [22]: Pure time-series model separating seasonal and trend components from a causal perspective to improve robustness.
- ST-Norm [6]: Introduces temporal and spatial normalization to refine high-frequency and local components.
- STWA [11]: Decomposes traffic into stable trends and fluctuating events, models them separately, and fuses predictions.
- SCNN [7]: Iteratively decomposes inputs into long-term, seasonal, short-term, and co-evolutionary components and fuses results.

• Methods considering distribution shift:

- AdaRNN [10]: Pure sequence model addressing distribution shift via clustering and dynamic context matching.
- CIGA [4]: Graph learning model capturing invariances via a causal view to generalize under shifts.
- STNSCM [8]: Neuralizes SCM and incorporates external conditions (e.g., temporal factors); here only temporal factors are used for fairness.
- CauSTG [26]: Enables generalization to shifted data by capturing invariant relationships.
- CaST [24]: Tackles spatio-temporal distribution shift via backdoor adjustment (temporal) and edge-level graph convolution (spatial).

5.4.1.3 Model Implementation Details

STEVE was implemented based on PyTorch 1.10.2 on an Ubuntu server equipped with an NVIDIA RTX 3090. In the TSRL module, the sizes of both temporal and spatial convolution kernels were set to 3. The hidden dimension D was searched within $\{16, 32, 64, 128\}$, the number of basic confounders K was selected from the range $\{16, 32, 64, 128, 256\}$, and the momentum coefficient γ in the confounder extractor was tested between 0.1 and 0.9. The model was trained using the Adam optimizer with a learning rate set to 0.001. The task balancing coefficients $\gamma_1, \gamma_2, \gamma_3$ were optimized using a dynamic weighted averaging strategy [16], with their initial values all set to 1.0.

Because data for unknown confounders is unavailable, simulated environments are constructed to assess the model's robustness to distributional shifts. Specifically, two types of shifts commonly observed in real-world settings are considered:

• *Temporal Distribution Shift (TDS)*: The temporal distribution is divided into weekdays and holidays. In the original training and test sets, the ratio of these two is approximately 5:2. To maximally simulate TDS, the experiments keep the training set unchanged and artificially set this ratio in the test set to 1:0 and 0:1, to examine the model's performance under extreme temporal distribution shifts.

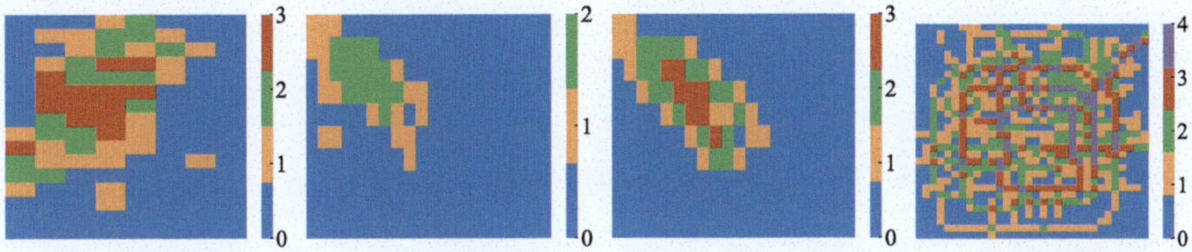

(a) NYCBike1 (b) NYCBike2 (c) NYCTaxi (d) BJTaxi

Fig. 5.6 Visualization of spatial clustering results

- *Spatial Distribution Shift (SDS)*: To simulate the spatial distribution characteristics of traffic entities in real-world scenarios, the k-means clustering algorithm is used to partition traffic entities into different clusters (as shown in Fig. 5.6), where cluster IDs are indicated by the color bar. The input to the clustering algorithm is the statistical information of traffic states, namely the mean, median, and standard deviation of historical traffic flow in the regions. The clustering results are denoted as $\{c_0, \ldots, c_k\}$, where clusters with smaller IDs usually correspond to non-core areas with lower traffic flow. The training and test sets initially have a mixed distribution of all clusters. Then, by controlling the distribution structure of the test set to retain only a single cluster, the SDS setting is constructed.

5.4.2 Overall Result Comparison and Analysis

All models in the experiments were run five times, and the average performance of each prediction method is reported in Tables 5.2, 5.3, and 5.4. In these tables, **bold** and underline indicate the best and second-best results, respectively. "Work" stands for weekdays, and "Holi" for holidays (i.e., non-weekdays), which are the tasks under the TDS scenario. c_i denotes the spatial entity cluster with ID i, representing tasks under the SDS scenario. "Avg" represents the average result across all tasks in the different scenarios. Based on the experimental results in the tables, the following observations can be drawn.

Across all four datasets and all distribution-shift settings (TDS/SDS), STEVE consistently ranks first, while the runner-up varies by task, indicating stronger robustness, stability, and adaptability than competing methods.

Unsupervised disentanglement methods yield no significant gains over classical spatiotemporal predictors; without supervisory signals they struggle to separate latent factors, so STEVE couples disentanglement with self-supervised signals.

Models tailored for distribution shift do not achieve the best results: AdaRNN and CIGA under-capture spatiotemporal dependencies; CauSTG emphasizes confounder-invariant relations while overlooking variable relations crucial for forecasting; and CaST enforces discretization that disrupts continuous-time structure and increases

Table 5.2 Comparison of results of different prediction methods under the temporal distribution shift scenario

Model	Dataset	NYCBike1			NYCBike2			NYCTaxi			BJTaxi		
	Task	Work	Holi	Avg	Work	Holi	Avg	Work	Holi	Avg	Work	Holi	Avg
STGCN	MAE	5.50	5.16	5.33	5.43	5.53	5.48	11.38	11.32	11.35	12.52	11.77	12.14
	MAPE	25.28	29.98	27.63	25.09	30.71	27.90	18.90	18.69	18.80	14.91	19.34	17.13
AGCRN	MAE	5.44	5.06	5.25	5.35	5.43	5.39	10.87	10.91	10.89	11.99	11.11	11.55
	MAPE	25.19	29.71	27.45	24.62	30.15	27.39	18.28	17.99	18.14	14.68	18.98	16.83
ASTGNN	MAE	5.69	5.31	5.50	5.25	5.62	5.43	10.99	11.28	11.13	12.09	11.03	11.56
	MAPE	25.34	28.82	27.08	28.16	35.24	31.70	19.45	24.27	21.86	15.31	19.77	17.54
HimNet	MAE	5.98	5.48	5.73	5.31	5.67	5.49	13.42	13.16	13.29	12.60	11.48	12.04
	MAPE	26.28	29.97	28.13	24.49	31.44	27.96	20.78	20.23	20.50	14.45	18.98	16.72
COST	MAE	6.64	6.34	6.49	7.06	7.50	7.28	13.14	13.02	13.08	14.05	13.87	13.96
	MAPE	29.67	37.62	33.65	31.23	39.32	35.28	32.80	30.37	31.59	17.10	22.41	19.76
ST-Norm	MAE	5.46	5.48	5.47	5.57	5.39	5.48	16.57	17.13	16.85	13.26	13.36	13.31
	MAPE	25.46	26.45	25.96	26.25	27.62	26.94	31.47	30.55	31.01	16.75	18.97	17.86
STWA	MAE	6.90	6.54	6.72	10.00	8.87	9.44	14.13	14.58	14.35	13.24	13.25	13.25
	MAPE	28.70	32.11	30.41	36.38	51.83	44.11	21.06	21.08	21.07	15.52	21.87	18.69
SCNN	MAE	6.55	6.16	6.35	5.78	5.65	5.71	12.62	12.78	12.70	12.68	11.80	12.24
	MAPE	27.49	32.90	30.20	25.99	31.24	28.62	21.29	20.88	21.09	15.24	19.37	17.31
AdaRNN	MAE	7.22	6.13	6.68	8.18	7.35	5.96	15.16	16.96	16.06	19.63	17.78	18.75
	MAPE	29.64	33.34	31.49	36.54	28.47	29.97	41.49	32.21	36.85	21.89	28.79	25.34
CIGA	MAE	6.47	5.76	6.12	6.05	5.86	5.96	15.23	15.34	15.29	13.47	12.69	13.08
	MAPE	29.27	32.61	30.94	31.49	28.45	29.97	18.95	21.51	20.23	16.71	22.24	19.48
STNSCM	MAE	5.97	5.29	5.63	6.15	5.76	5.96	14.69	14.95	14.82	13.80	11.29	12.55
	MAPE	26.67	29.91	28.29	27.88	31.13	29.51	23.63	23.39	23.51	16.96	19.36	18.16
CauSTG	MAE	8.01	5.67	6.84	7.26	5.50	6.38	16.08	15.61	15.85	17.31	25.93	21.62
	MAPE	29.86	29.53	29.70	28.87	28.72	28.80	31.98	30.22	31.10	19.27	30.15	24.71
CaST	MAE	5.70	5.99	5.84	6.25	7.19	6.72	13.15	14.36	13.76	12.93	11.76	12.35
	MAPE	26.31	29.43	27.87	28.60	36.19	32.40	19.53	18.44	18.99	16.45	21.29	18.87
STEVE	MAE	**5.13**	**4.73**	**4.93**	**4.75**	**4.98**	**4.87**	**10.43**	**10.51**	**10.47**	**11.22**	**10.66**	**10.94**
	MAPE	**22.89**	**26.13**	**24.51**	**20.57**	**25.31**	**22.94**	**16.46**	**16.06**	**16.26**	**13.97**	**18.95**	**16.46**

modeling complexity. These observations support our basis-vector approach to dynamic confounders without discretization.

Comparing bicycle and taxi data under TDS (Table 5.2) and SDS (Tables 5.3 and 5.4) shows STEVE's advantage is more pronounced on bicycle datasets, likely because bicycle travel is more sensitive to confounders and exhibits larger spatiotemporal shifts; this is further corroborated in Sect. 5.4.4.2.

Although baselines such as AGCRN and STNSCM sometimes approach STEVE in MAE, a sizable gap remains in MAPE, indicating that STEVE achieves lower relative error and thus better cross-scenario generalization.

To further verify the significant performance improvement of the proposed model compared to the baseline models, the Nemenyi significance test was employed, and a Critical Difference (CD) diagram was plotted for analysis. As shown in Fig. 5.7, the horizontal axis represents the average performance ranking of each model on both MAE and MAPE metrics. When the ranking difference between two models is less than the CD value (at a 5% significance level), they are connected by a thick

Table 5.3 Comparison of results of different prediction methods on bicycle datasets under the spatial distribution shift scenario

Model	Dataset	NYCBike1					NYCBike2			
	Task	c_0	c_1	c_2	c_3	Avg	c_0	c_1	c_2	Avg
STGCN	MAE	2.96	4.36	5.81	7.53	5.16	3.94	4.90	7.05	5.30
	MAPE	33.86	28.50	24.56	22.28	27.30	37.47	26.47	20.17	28.04
AGCRN	MAE	2.93	4.35	5.87	7.56	5.18	3.80	4.74	7.00	5.18
	MAPE	33.46	28.88	25.41	22.61	27.59	36.73	26.04	19.75	27.51
ASTGNN	MAE	3.40	4.72	6.38	8.21	5.68	3.58	4.94	7.34	5.29
	MAPE	34.18	28.55	25.19	22.93	27.71	37.62	27.86	22.06	29.18
HimNet	MAE	3.18	4.63	6.15	8.07	5.51	3.95	4.75	7.11	5.27
	MAPE	31.98	28.84	26.00	23.88	27.68	36.50	25.87	20.61	27.66
COST	MAE	3.13	4.98	7.28	9.24	6.16	3.91	5.87	10.38	6.72
	MAPE	31.43	32.16	33.65	29.67	31.73	36.68	31.92	33.54	34.05
ST-Norm	MAE	3.40	4.42	5.76	7.98	5.39	3.46	5.04	7.01	5.17
	MAPE	33.29	27.42	24.31	21.26	26.57	33.33	28.86	21.42	27.87
STWA	MAE	3.73	5.37	7.27	9.46	6.46	3.86	7.77	15.62	9.08
	MAPE	37.60	31.66	27.70	25.65	30.65	34.07	44.37	44.06	40.83
SCNN	MAE	3.48	5.13	7.03	8.76	6.10	3.64	5.15	7.82	5.54
	MAPE	34.73	31.10	27.90	24.84	29.64	34.60	28.01	22.70	28.44
AdaRNN	MAE	3.27	5.45	8.00	10.35	6.77	5.02	7.56	13.53	8.71
	MAPE	30.80	32.83	31.57	28.52	30.93	39.72	38.49	40.66	39.62
CIGA	MAE	3.21	4.73	6.72	9.07	5.93	3.31	5.64	9.56	6.17
	MAPE	34.82	28.69	25.08	21.99	27.64	37.79	28.77	29.34	31.97
STNSCM	MAE	3.03	4.60	5.86	8.37	5.47	6.11	6.71	6.82	6.54
	MAPE	26.84	27.00	23.28	20.24	24.34	29.68	27.62	27.68	28.33
CauSTG	MAE	4.07	6.00	7.97	9.61	6.91	4.30	6.46	10.78	7.18
	MAPE	38.46	32.83	27.33	24.12	30.69	38.22	30.75	25.85	31.61
CaST	MAE	2.71	4.75	6.18	8.15	5.45	5.01	5.70	8.59	6.43
	MAPE	26.80	29.77	26.91	24.69	27.04	45.59	27.31	24.07	32.32
STEVE	MAE	**2.21**	**4.13**	**5.59**	**7.09**	**4.75**	**2.82**	**4.46**	**6.64**	**4.64**
	MAPE	**24.44**	**26.89**	**23.17**	**20.06**	**23.64**	**24.72**	**23.61**	**18.88**	**22.40**

black line, indicating that their difference is not statistically significant. Otherwise, there is a significant performance gap between them. From the results in the figure, it can be observed that the proposed STEVE model is significantly better than the best baseline model at the 5% significance level, while the performance differences among the baseline models are not significant.

5.4.3 Ablation Study of Key Modules

This section mainly analyzes the rationality of the model design and the sensitivity of hyperparameters.

Table 5.4 Comparison of results of different prediction methods on taxi datasets under the spatial distribution shift scenario

Model	Dataset	NYCTaxi					BJTaxi					
	Task	c_0	c_1	c_2	c_3	Avg	c_0	c_1	c_2	c_3	c_4	Avg
STGCN	MAE	3.97	8.31	17.17	27.56	14.25	4.97	9.45	13.84	20.64	30.13	15.80
	MAPE	27.68	16.80	11.42	9.74	16.41	23.88	15.41	12.38	10.56	9.34	14.31
AGCRN	MAE	3.77	8.17	17.12	27.68	14.19	5.93	11.16	20.51	21.71	31.44	18.15
	MAPE	26.14	16.76	11.34	9.65	15.97	28.87	15.37	12.77	11.46	9.29	15.55
ASTGNN	MAE	4.35	8.50	17.64	27.87	14.59	4.95	9.37	13.40	19.74	27.89	15.07
	MAPE	33.39	17.44	11.48	9.63	17.98	22.99	15.06	12.04	10.90	9.53	14.08
HimNet	MAE	4.36	9.59	20.25	33.64	16.96	5.09	9.62	14.18	20.84	30.69	16.08
	MAPE	29.18	18.80	13.16	10.98	18.03	22.49	15.42	12.59	10.62	9.43	14.11
COST	MAE	4.70	15.10	38.92	67.02	31.44	5.18	10.35	15.75	24.80	37.31	18.68
	MAPE	31.11	33.55	34.19	33.07	32.98	23.76	18.62	15.91	14.50	12.94	17.15
ST-Norm	MAE	6.01	13.09	23.92	39.05	20.51	5.79	10.71	15.50	22.19	31.08	17.05
	MAPE	45.77	30.53	18.23	16.37	27.72	25.63	16.87	13.38	10.79	8.99	15.13
STWA	MAE	4.09	9.97	23.45	37.35	18.72	5.22	10.16	15.30	23.09	33.98	17.55
	MAPE	28.11	19.79	15.24	13.09	19.05	23.99	17.13	14.27	12.19	10.71	15.66
SCNN	MAE	4.19	9.26	20.27	31.58	16.33	5.25	9.82	14.39	21.16	30.05	16.13
	MAPE	29.29	19.45	14.49	12.17	18.85	23.67	16.00	13.11	11.00	9.49	14.65
AdaRNN	MAE	4.81	17.97	29.20	35.25	21.81	6.33	13.99	22.63	33.67	52.23	25.77
	MAPE	35.52	35.59	40.89	46.73	39.65	27.86	24.17	22.03	20.02	17.50	22.32
CIGA	MAE	5.32	8.76	19.19	27.65	15.23	7.81	12.09	16.29	22.01	31.03	17.85
	MAPE	27.11	17.02	13.33	16.09	18.39	29.14	16.88	13.91	12.99	10.58	16.70
STNSCM	MAE	4.75	8.54	17.72	25.69	13.83	6.79	10.87	14.09	21.96	29.23	16.59
	MAPE	24.11	22.65	12.43	11.11	17.58	23.87	18.43	14.22	12.22	9.64	15.67
CauSTG	MAE	6.95	15.85	36.19	65.90	31.22	6.35	13.51	22.75	37.88	58.99	27.90
	MAPE	42.07	29.67	22.23	19.71	28.42	28.44	20.77	19.69	18.95	18.07	21.18
CaST	MAE	4.89	9.87	16.90	26.91	14.64	5.28	10.11	14.42	21.12	31.25	16.44
	MAPE	31.84	20.30	14.55	11.84	19.63	24.22	17.41	14.35	11.04	10.47	15.50
STEVE	MAE	**3.47**	**8.02**	**16.52**	**25.61**	**13.40**	**4.62**	**8.90**	**12.94**	**18.91**	**27.00**	**14.47**
	MAPE	**21.93**	**16.20**	**11.08**	**9.36**	**14.64**	**22.19**	**15.05**	**12.03**	**10.11**	**8.65**	**13.61**

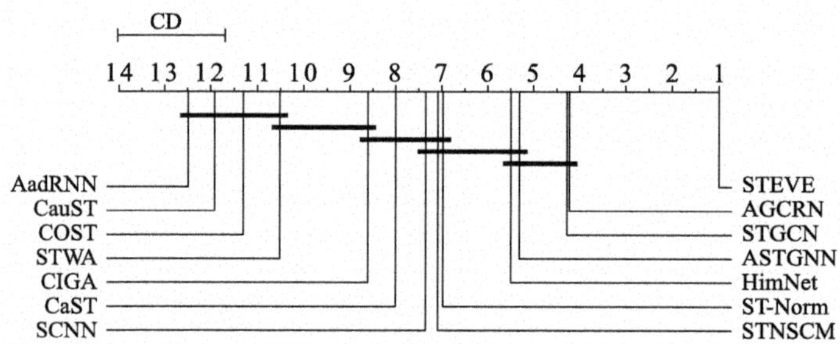

Fig. 5.7 Critical difference diagram for the significance test of different methods

Table 5.5 Ablation study results for the STEVE model

Dataset	NYCTaxi		NYCBike1		NYCBike2		BJTaxi	
Metric	TDS	SDS	TDS	SDS	TDS	SDS	TDS	SDS
STEVE	**10.47**	**13.40**	**4.93**	**4.75**	**4.87**	**4.64**	**10.94**	**14.47**
(i) *w/o cfd*	10.65	13.67	5.04	4.85	5.04	4.79	11.13	14.81
(ii) *w/o ssl*	11.70	15.01	4.96	4.77	5.06	4.94	11.78	15.56
(iii) *w/o ad*	10.54	13.52	4.97	4.79	4.98	4.75	11.11	14.73
(iv) *w/o mi*	10.72	13.69	4.96	4.78	5.03	4.80	11.00	14.55

5.4.3.1 Module Ablation Analysis

To analyze the impact of each component proposed in the STEVE model on the final prediction performance, an ablation study was conducted in this section.

Specifically, the following model variants were constructed to analyze the influence of different modules: (i) *w/o cfd*: Removes the confounder bank and directly uses $\boldsymbol{Q}$ as the confounder representation. (ii) *w/o ssl*: Removes the self-supervised learning (SSL) tasks in (5.21). (iii) *w/o ad*: Disables the adversarial disentanglement module in (5.24). (iv) *w/o mi*: Does not use the mutual information regularization in (5.26). The average MAE results for all datasets are shown in Table 5.5.

The experimental results show that all four components contribute to the overall model performance. Specifically, the performance degradation of variants (i) and (ii) is particularly significant, indicating that the proposed confounder bank, guided by self-supervised learning, can effectively inject information from representative confounders and extract reasonable confounder representations from limited observable spatio-temporal data. Furthermore, the removal of these components has a more pronounced impact on the NYCTaxi and BJTaxi tasks compared to NYCBike1 and NYCBike2. This may be because the spatio-temporal relationships in taxi data are more complex, making them more sensitive to confounder modeling, thus amplifying the performance drop caused by component removal.

5.4.3.2 Hyperparameter Impact Analysis

The hyperparameter sensitivity analysis examines the momentum coefficient γ, the number of confounder basis vectors K, and the representation dimension D. Results are summarized in Fig. 5.8.

First, Fig. 5.8a reports the effect of γ across datasets. Here γ varies from 0.1 to 0.9 (with intermediate values omitted for clarity). The optimal settings are: NYCTaxi

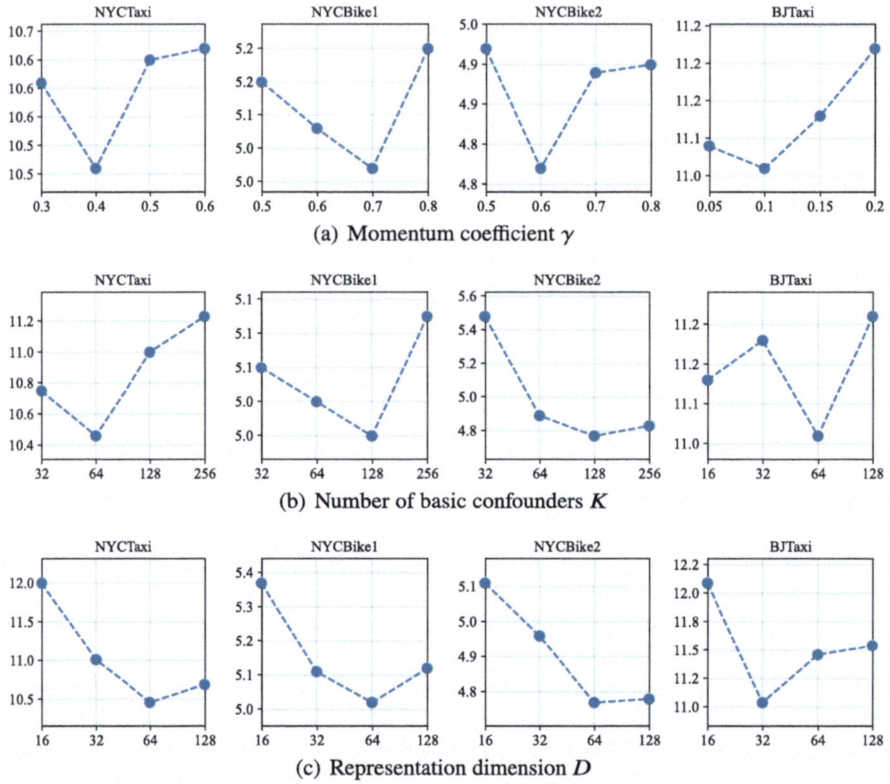

Fig. 5.8 Hyperparameter analysis results for the STEVE model

at $\gamma = 0.4$; NYCBike1 at $\gamma = 0.7$; NYCBike2 at $\gamma = 0.6$; and BJTaxi at $\gamma = 0.1$. These differences may reflect dataset-specific degrees of confounder influence.

Second, Fig. 5.8b shows the impact of K. The best performance is obtained with $K = 64$ for NYCTaxi and BJTaxi, and with $K = 128$ for NYCBike1 and NYCBike2.

Finally, Fig. 5.8c evaluates the representation dimension $D \in \{16, 32, 64, 128\}$. The setting $D = 64$ is suitable for NYCTaxi, NYCBike1, and NYCBike2, whereas BJTaxi performs best at $D = 32$. These patterns suggest heterogeneous spatio-temporal dependencies across datasets, supporting the use of dataset-specific representation dimensions.

5.4.4 In-Depth Analysis of Model Behavior

This subsection conducts an in-depth analysis of model behavior from multiple perspectives. It examines the effectiveness of confounder learning, evaluates generalization performance under unseen confounders, analyzes learned representations

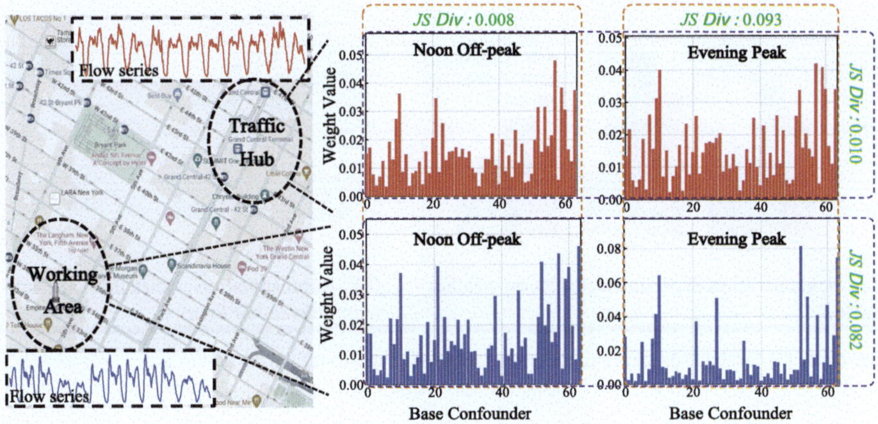

Fig. 5.9 Confounder distributions at different locations and times

through visualization, and assesses model efficiency and scalability, providing comprehensive insights into robustness and practical deployment.

5.4.4.1 Analysis of Confounder Learning Effectiveness

As described in Sect. 5.3.1, the proposed confounder extractor can generate meaningful confounder representations for each sample based on a unique weight vector $\boldsymbol{\alpha}$ and the basic confounder bank.

To verify the effectiveness of this method, this part of the experiment visualizes the weight distributions of different samples from the NYCTaxi dataset in Fig. 5.9, where JS divergence denotes the Jensen-Shannon divergence.[2]

Through an in-depth analysis of the experimental results, two phenomena can be observed: First, there is a significant difference in the weight distributions between work areas and traffic hubs (especially during peak hours), indicating that the model can learn environmental features associated with these two types of areas.

Second, from the non-peak hours at noon to the evening peak hours, the weight distribution of work areas gradually concentrates on a few primary confounders, while the distribution of traffic hubs remains relatively dispersed. This suggests that the external environment of traffic hubs is consistently complex, whereas work areas exhibit stronger regularity during the evening peak, further validating the model's effective capture of features from different regions.

[2] JS divergence: https://en.wikipedia.org/wiki/Jensen%E2%80%93Shannon_divergence.

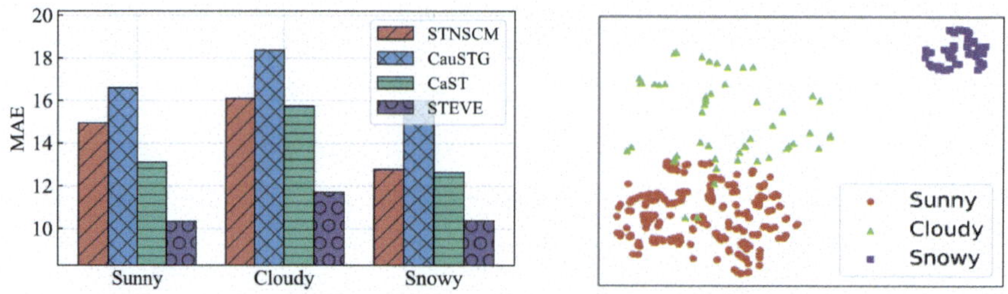

(a) Generalization experiment results on weather factors

(b) Representation distribution of learned weather factors

Fig. 5.10 Generalization experiment results on unseen confounders on taxi data

5.4.4.2 Generalization Analysis on Unseen Confounders

In the COSSL module in Sect. 5.3.2, representative self-supervised signals were selected to guide the model in learning information related to latent confounders, thereby enhancing its robustness and generalization ability in different environments.

To evaluate the effectiveness of this method, this experiment collected NYCTaxi weather data, which was not used in model training, and compared STEVE with three spatio-temporal models that address distribution shifts without using self-supervised signals.

The experimental results are shown in Fig. 5.10a. The results show that STEVE significantly outperforms the other baseline models under all three weather conditions. This validates the effectiveness of self-supervised signals in enhancing confounder representation learning, enabling the model to adapt more robustly to unseen confounders.

Furthermore, we analyze the learned weather confounder representations and visualize them using the *t-SNE* algorithm [17] (Fig. 5.10b). According to the results, we find two observations. First, embeddings under the same weather condition form relatively compact clusters; in particular, snowy-day representations are tighter than sunny-day ones, indicating higher homogeneity under snow. Second, in the representation space, the sunny cluster lies closer to cloudy and farther from snowy, aligning with real-world semantics. These findings suggest that STEVE, guided by self-supervised signals, learns information-rich confounder representations consistent with environmental characteristics.

Across all weather conditions, STEVE consistently outperforms baseline models, indicating that self-supervised signals enhance confounder representation learning and improve robustness to unseen confounders.

Because bicycle travel is generally more sensitive to weather changes than taxi demand, we additionally collect weather data for the NYCBike1 dataset and compare STEVE against causality-based spatiotemporal prediction baselines. The results in Table 5.6 summarize model performance. The average error reduction rate refers

Table 5.6 Generalization experiment results on unseen confounders on bicycle data

Model	STNSCM		CauSTG		CaST		STEVE		Avg. Error reduction rate	
Metric	MAE	MAPE	MAE	MAPE	MAE	MAPE	MAE	MAPE	MAE (%)	MAPE (%
Sunny	4.93	23.91	5.04	24.10	4.91	23.94	**4.85**	**23.29**	1.69	2.59
Cloudy	5.90	28.43	6.25	29.95	5.68	26.79	**5.09**	**24.77**	13.76	12.87
Snowy	7.09	32.17	8.41	34.30	6.26	29.67	**5.21**	**25.23**	26.51	21.57

to the mean percentage decrease in prediction error achieved by STEVE relative to competing baselines.

There are three phenomena by observing the results in the table. First, the proposed STEVE model outperforms other baseline models on both metrics under all weather conditions, achieving the lowest prediction error. This demonstrates that the proposed basis vector method has sufficient expressive power to effectively model arbitrary confounders, even if they did not appear during model training.

Second, compared to the taxi data in Fig. 5.10a, the prediction errors for the bicycle dataset differ more significantly across various weather conditions, indicating that bicycle travel is more sensitive to weather changes. A horizontal comparison of prediction results across different weather conditions shows that the prediction errors for sunny and cloudy days are similar, while the prediction error for snowy days is significantly higher than for the other two conditions. This suggests that extreme weather, like snow, has the greatest impact on bicycle travel.

Third, under extreme weather conditions such as snow, unlike other baseline models, the prediction performance of the STEVE model does not degrade significantly. This indicates that the proposed STEVE method has excellent generalization performance under extreme conditions, offering a potential solution to the problem presented in Fig. 1.1b of the first chapter.

5.4.4.3 Representation Visualization Analysis

In Sect. 5.3.3, the Mutual Information Minimization (MIM) loss function was proposed to disentangle the confounder-irrelevant representation $\boldsymbol{H}$ and the confounder representation $\boldsymbol{C}$ from a distributional perspective.

To verify the effectiveness of this method, this experiment randomly samples from the test set to generate corresponding $\boldsymbol{H}$ and $\boldsymbol{C}$ representations, and uses the t-SNE algorithm to reduce their dimensionality to two for visualization. The visualization results are shown in Fig. 5.11, where S represents the Silhouette Score [20], used to measure clustering effectiveness. A higher value indicates better clustering performance, and vice versa.

By comparing and analyzing the visualization results, it can be found that before introducing the MIM loss, the distributions of $\boldsymbol{H}$ and $\boldsymbol{C}$ in the representation space are mixed, making them difficult to distinguish. After introducing the MIM loss, the

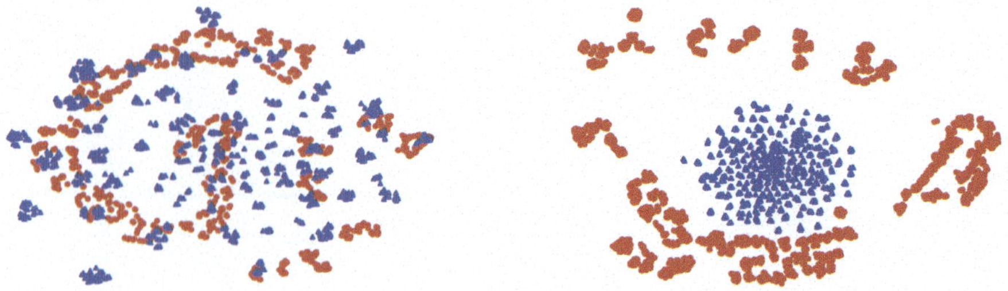

(a) Without MIM ($S = 0.006$) (b) With MIM ($S = 0.188$)

Fig. 5.11 Visualization of confounder-relevant representations (red circles) and irrelevant representations (purple triangles)

distributions of $\boldsymbol{H}$ and $\boldsymbol{C}$ are clearly more separated. All $\boldsymbol{H}$ representations form one cluster, and all $\boldsymbol{C}$ representations form another, with a much clearer distinction between them. The silhouette score also increases from 0.006 to 0.188, indicating that the clustering quality of the learned representations has improved. Based on these results, it can be concluded that the MIM loss can effectively enhance the distributional independence of the disentangled representations.

5.4.4.4 Model Efficiency and Scalability Analysis

This section evaluates the model's efficiency and scalability.

First, for efficiency, training and inference times per epoch are measured for all methods on all datasets. The results are summarized in Table 5.7. To ensure comparability, all experiments are conducted under identical conditions—an Ubuntu server with an NVIDIA RTX 3090—and the batch size is held constant.

The experimental results show that, compared to the best-performing baseline AGCRN, the proposed model's training and inference times are reduced by 73.7 and 81.9%, respectively. Although some baselines like STGCN and COST have lower time costs than our model, when considering the overall performance results in Tables 5.2, 5.3, and 5.4, it is evident that our model significantly improves training efficiency while ensuring high prediction accuracy, achieving a good balance between predictive performance and computational overhead.

Next, we explore the scalability of STEVE compared to AGCRN (the best-performing baseline model), focusing on its ability to handle changes in dataset size and graph structure. The experiments are based on the BJTaxi dataset, which covers traffic data from 1024 graph nodes over 4 months. The experimental results are shown in Fig. 5.12. In terms of dataset size, 25% represents one month of data, 50% represents two months, and so on. In terms of graph structure, the original graph is divided into four connected subgraphs with an equal number of nodes. 25% means using only the nodes from the first subgraph to build the adjacency matrix, 50% represents nodes from the first two subgraphs, and so on.

Table 5.7 Comparison of training/inference time of different prediction methods (Unit: seconds/epoch)

Dataset	NYCTaxi	NYCBike1	NYCBike2	BJTaxi
STGCN	1.39/0.05	1.07/0.07	1.37/0.06	26.22/1.39
AGCRN	25.94/6.34	9.82/1.24	25.23/5.95	68.22/13.05
ASTGNN	12.91/1.14	9.48/0.74	10.14/0.96	35.16/3.43
HimNet	5.43/1.27	9.26/1.96	5.63/1.28	53.37/9.63
COST	3.21/1.42	3.41/1.82	3.27/1.37	6.32/5.98
ST-Norm	5.28/0.37	9.62/0.48	5.02/0.36	12.98/0.78
STWA	13.40/2.43	13.62/1.32	12.53/2.23	53.61/10.90
SCNN	15.34/3.54	14.96/1.91	14.72/2.74	65.03/7.41
AdaRNN	16.92/2.43	8.83/1.13	16.32/2.01	55.78/8.26
CIGA	17.36/0.95	16.73/0.67	16.89/0.87	61.06/3.34
STNSCM	1.66/0.19	2.79/0.32	1.74/0.19	4.94/0.68
CauSTG	3.22/1.61	4.92/2.33	3.15/1.57	10.62/4.30
CaST	25.31/7.20	37.82/8.13	21.07/6.52	81.93/18.73
STEVE	4.51/0.52	3.83/0.45	2.22/0.31	27.13/2.98

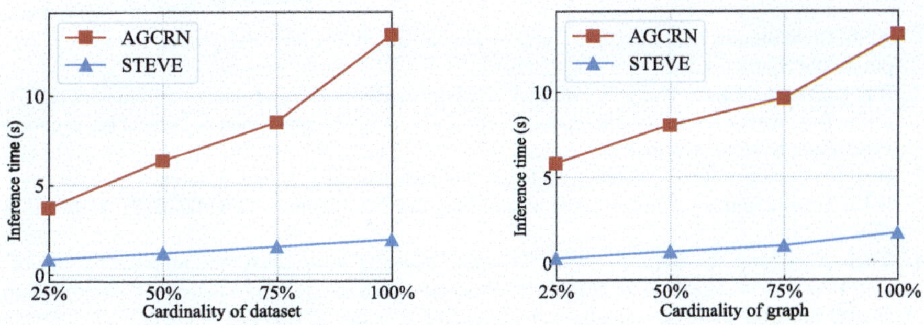

Fig. 5.12 Scalability experiment results for the STEVE model

Figure 5.12 shows that prediction time for both methods increases with dataset size and graph size. However, relative to the steep growth observed for AGCRN, STEVE exhibits a markedly flatter scaling trend. These results substantiate the superior scalability of the proposed model in large-scale spatio-temporal forecasting tasks.

5.5 Summary

This chapter proposes the STEVE model, which represents a novel attempt to extend the backdoor adjustment method to handle continuous or undefined confounders in spatio-temporal forecasting. The model creates a basic confounder bank, representing any unknown confounder as an adaptive linear combination of a set of basic confounder representations to express a wide variety of different confounders. This approach effectively extends the backdoor adjustment method in causal modeling, enabling the model to learn stable causal relationships even in the presence of continuous confounders. Next, the STEVE model leverages three meticulously designed self-supervised auxiliary tasks—from temporal, spatial, and semantic perspectives—to thoroughly learn the representations of these basic confounders. Finally, mutual information minimization and adversarial disentanglement methods are employed to separately model the confounder-irrelevant relationship and the confounder effect. Based on the predictions for the future state from these two types of relationships, they are fused to produce the final forecast result.

Through systematic experiments on four traffic datasets, the robustness and adaptability of STEVE were validated. Furthermore, generalization experiments were conducted to demonstrate that STEVE can effectively learn valid representations of confounders not seen during the training process.

References

1. Abdi, H., Williams, L.J.: Principal component analysis. Wiley Interdisciplinary Reviews: Computational Statistics **2**(4), 433–459 (2010)
2. Bai, L., Yao, L., Li, C., Wang, X., Wang, C.: Adaptive graph convolutional recurrent network for traffic forecasting. In: Proceedings of the 34th International Conference on Neural Information Processing Systems (NeurIPS'20), pp. 17804–17815 (2020)
3. Bengio, Y., Courville, A., Vincent, P.: Representation learning: A review and new perspectives. IEEE Transactions on Pattern Analysis and Machine Intelligence (TPAMI) **35**(8), 1798–1828 (2013)
4. Chen, Y., Zhang, Y., Bian, Y., et al.: Learning causally invariant representations for out-of-distribution generalization on graphs. In: Proceedings of the 36th International Conference on Neural Information Processing Systems (NeurIPS'22), pp. 22131–22148 (2022)
5. Cheng, P., Hao, W., Dai, S., Liu, J., Gan, Z., Carin, L.: CLUB: A contrastive log-ratio upper bound of mutual information. In: Proceedings of the 37th International Conference on Machine Learning (ICML'20), pp. 1779–1788 (2020)
6. Deng, J., Chen, X., Jiang, R., Song, X., Tsang, I.W.: ST-Norm: Spatial and temporal normalization for multi-variate time series forecasting. In: Proceedings of the 27th ACM SIGKDD Conference on Knowledge Discovery and Data Mining (KDD'21), pp. 269–278 (2021)
7. Deng, J., Chen, X., Jiang, R., Yin, D., Yang, Y., Song, X., Tsang, I.W.: Disentangling structured components: Towards adaptive, interpretable and scalable time series forecasting. IEEE Transactions on Knowledge and Data Engineering (TKDE) **36**(8), 3783–3800 (2024)
8. Deng, P., Zhao, Y., Liu, J., Jia, X., Wang, M.: Spatio-temporal neural structural causal models for bike flow prediction. In: Proceedings of the 37th AAAI Conference on Artificial Intelligence (AAAI'23), pp. 4242–4249 (2023)
9. Dong, Z., Jiang, R., Gao, H., Liu, H., Deng, J., Wen, Q., Song, X.: Heterogeneity-informed meta-parameter learning for spatiotemporal time series forecasting. In: Proceedings of the 30th ACM SIGKDD Conference on Knowledge Discovery and Data Mining (KDD'24), pp. 631–641 (2024)
10. Du, Y., Wang, J., Feng, W., Pan, S., Qin, T., Xu, R., Wang, C.: AdaRNN: Adaptive learning and forecasting of time series. In: Proceedings of the 30th ACM International Conference on Information and Knowledge Management (CIKM'21), pp. 402–411 (2021)

11. Fang, Y., Qin, Y., Luo, H., Zhao, F., Xu, B., Zeng, L., Wang, C.: When spatio-temporal meet wavelets: Disentangled traffic forecasting via efficient spectral graph attention networks. In: Proceedings of the 2023 IEEE 39th International Conference on Data Engineering (ICDE'23), pp. 517–529 (2023)
12. Ganin, Y., Lempitsky, V.S.: Unsupervised domain adaptation by backpropagation. In: Proceedings of the 32nd International Conference on Machine Learning (ICML'15), pp. 1180–1189 (2015)
13. Guo, S., Lin, Y., Wan, H., Li, X., Cong, G.: Learning dynamics and heterogeneity of spatial-temporal graph data for traffic forecasting. IEEE Transactions on Knowledge and Data Engineering (TKDE) **34**(11), 5415–5428 (2022)
14. Ji, J., Wang, J., Huang, C., Wu, J., Xu, B., Wu, Z., Junbo, Z., Zheng, Y.: Spatio-temporal self-supervised learning for traffic flow prediction. In: Proceedings of the 37th AAAI Conference on Artificial Intelligence (AAAI'23), pp. 4356–4364 (2023)
15. Ji, J., Zhang, W., Wang, J., Huang, C.: Seeing the unseen: Learning basis confounder representations for robust traffic prediction. In: Proceedings of the 31st ACM SIGKDD Conference on Knowledge Discovery and Data Mining (KDD'25), pp. 1–12 (2025)
16. Liu, S., Johns, E., Davison, A.J.: End-to-end multi-task learning with attention. In: Proceedings of the 2019 IEEE Conference on Computer Vision and Pattern Recognition (CVPR'19), pp. 1871–1880 (2019)
17. Maaten, L.V.D., Hinton, G.: Visualizing data using t-SNE. Journal of Machine Learning Research (JMLR) **9**(86), 2579–2605 (2008)
18. Pearl, J.: Causality: Models, Reasoning and Inference. Cambridge University Press (2000)
19. Reynolds, D.A.: Gaussian mixture models. Encyclopedia of Biometrics **741**(3), 659–663 (2009)
20. Rousseeuw, P.J.: Silhouettes: A graphical aid to the interpretation and validation of cluster analysis. Journal of Computational and Applied Mathematics **20**(1), 53–65 (1987)
21. Vaswani, A., Shazeer, N., Parmar, N., Uszkoreit, J., Jones, L., Gomez, A.N., Kaiser, L., Polosukhin, I.: Attention is all you need. In: Proceedings of the 31st International Conference on Neural Information Processing Systems (NeurIPS'17), pp. 5998–6008 (2017)
22. Woo, G., Liu, C., Sahoo, D., Kumar, A., Hoi, S.: CoST: Contrastive learning of disentangled seasonal-trend representations for time series forecasting. In: Proceedings of the 10th International Conference on Learning Representations (ICLR'22), pp. 1–18 (2022)
23. Wu, Z., Pan, S., Long, G., Jiang, J., Zhang, C.: Graph WaveNet for deep spatial-temporal graph modeling. In: Proceedings of the 28th International Joint Conference on Artificial Intelligence (IJCAI'19), pp. 1907–1913 (2019)
24. Xia, Y., Liang, Y., Wen, H., Liu, X., Wang, K., Zhou, Z., Zimmermann, R.: Deciphering spatio-temporal graph forecasting: A causal lens and treatment. In: Proceedings of the 37th International Conference on Neural Information Processing Systems (NeurIPS'23), pp. 37068–37088 (2023)
25. Yu, B., Yin, H., Zhu, Z.: Spatio-temporal graph convolutional networks: A deep learning framework for traffic forecasting. In: Proceedings of the 27th International Joint Conference on Artificial Intelligence (IJCAI'18), pp. 3634–3640 (2018)
26. Zhou, Z., Huang, Q., Yang, K., Wang, K., Wang, X., Zhang, Y., Liang, Y., Wang, Y.: Maintaining the status quo: Capturing invariant relations for OOD spatiotemporal learning. In: Proceedings of the 29th ACM SIGKDD Conference on Knowledge Discovery and Data Mining (KDD'23), pp. 3603–3614 (2023)

Chapter 6
Decomposition Prediction Framework

Abstract Due to the open nature of traffic environments, traffic state data is typically influenced by a confluence of factors, resulting in complex mixed distributions. Effective modeling of these mixed distributions in spatiotemporal data is crucial for enhancing the stability of intelligent transportation systems in complex scenarios. However, the performance upper bound of existing forecasting models is often constrained by the complexity of these spatiotemporal mixed distributions. To address this, this chapter introduces an information theory-based spatiotemporal decomposition forecasting framework. Grounded in theoretical guarantees, this model-agnostic framework reduces the complexity of input data through a decomposition algorithm, thereby overcoming the predictive performance bottlenecks inherent in various existing models.

6.1 Modeling Motivation

In urban traffic scenarios, the evolution of traffic flow is often driven by the joint action of multiple factors. These factors are intertwined, causing traffic data to exhibit characteristics of mixed distributions; that is, data distribution under different conditions follows not a single pattern, but a complex structure formed by the superposition, switching, or interaction of multiple latent distributions. If a traffic forecasting model fails to correctly model this mixed distribution characteristic, its prediction accuracy and generalization capability will be severely compromised when facing complex and dynamic traffic environments. Modeling mixed distributions can effectively enhance a model's adaptability to diverse traffic modes. For instance, during daily commuting hours, traffic flow is primarily driven by work-related travel demand, whereas during holidays, factors such as leisure travel and tourist flows may dominate the traffic pattern. Consequently, the same road segment may exhibit distinct data distributions during different time periods. Therefore, by modeling mixed distributions, models can better capture the latent structures of different traffic modes, thereby enabling accurate predictions across varying scenarios.

Despite significant progress in deep spatiotemporal forecasting models, most assume consistent spatial and temporal correlations across all geographical units.

J. Wang and J. Ji, *Mechanism-Driven Explainable Urban Spatio-Temporal Prediction*, Big Data Management, https://doi.org/10.1007/978-981-92-0662-9_6

Under this assumption, models learn a set of shared parameters from data across all units and apply them to predict traffic conditions at different locations and times. However, real-world traffic data is inherently complex and heterogeneous. Influenced by socioeconomic events such as commuting, entertainment, and commercial activities, human mobility patterns generate diverse spatial and temporal correlations. Numerous studies suggest that traffic data can be decomposed into multiple sub-components, each exhibiting relatively simple yet distinct spatial and temporal patterns [6, 27].

To address the aforementioned complexity, the "Decompose-Then-Predict" (DTP) paradigm has increasingly garnered attention in the field of traffic forecasting. This approach decomposes traffic data into distinct components and assigns independent parameters for modeling and prediction, thereby better capturing the heterogeneous characteristics of traffic data. Early research employed a two-stage approach: first utilizing traditional signal decomposition tools to preprocess traffic data into multiple components, and then predicting each component separately. Common decomposition tools include seasonal-trend decomposition [28] and wavelet transform [22, 26]. Recent studies have integrated decomposition techniques directly into deep neural networks, constructing end-to-end multi-component traffic forecasting models [5, 7, 21, 24]. Empirical results from existing research indicate that within the DTP framework, decomposition modules significantly enhance prediction performance. Although many studies have experimentally validated the potential of the DTP method in improving prediction accuracy, two fundamental questions regarding decomposition methods remain effectively unresolved.

Q1: What type of decomposition method guarantees that predictions based on decomposed components outperform predictions based on original data? Intuitively, the DTP paradigm adopts a "divide and conquer" strategy, breaking down the complex traffic prediction problem into multiple simpler, more manageable subproblems. However, not all decomposition methods effectively improve prediction outcomes. For example, if a traffic state sequence is decomposed into N components where each component's magnitude is $1/N$ of the original sequence, yet the length of each decomposed sequence remains identical to the original, the complexity of predicting each component may not be reduced. Therefore, rigorous theoretical analysis is required to identify decomposition methods that effectively enhance prediction performance.

Q2: Is it possible to construct a theoretically guided decomposition framework to improve the predictive performance of various traffic models? If a theoretical answer to Q1 is obtained, the next step is to develop a corresponding decomposition framework based on these theoretical insights. This framework must ensure that the decomposed components possess superior predictability compared to the original data. Furthermore, as a genuine framework, it should possess good universality and compatibility, capable of adapting to a variety of traffic prediction models.

To address these two problems, this chapter follows the noisy channel coding theory in information theory. Utilizing information-theoretic tools such as Fano's Inequality, we derive an effective decomposition theory capable of reducing data complexity. Based on constraints defined by regularization terms, we design a model-agnostic, portable spatiotemporal decomposition forecasting framework that breaks through the performance bottlenecks of various spatiotemporal models. Extensive experiments on real-world datasets demonstrate that this framework achieves an average performance improvement of up to 19.37% across various traffic models.

First, based on information theory, a set of theoretical analysis methods is proposed to answer Q1. This method abstracts the traffic prediction problem as a transmission and decoding process over a noisy channel, where historical traffic states are viewed as information transmitted through the channel, and the prediction function acts as the decoder for future traffic states. Under this framework, traffic prediction error is decomposed into two parts: error induced by data (transmission error) and error induced by the model (decoding error). Leveraging Fano's Inequality [8], a sufficient condition is derived to ensure that the decomposition algorithm results in a total data error for all sub-problems that is less than the data error of the original problem. This condition indicates that an effective decomposition algorithm should maximize the independence of decomposed components, a concept defined herein as the "Component Independence (CI) Principle."

Second, guided by the CI Principle, a novel Theory-guided Graph Decomposition Learning (TGDL) framework is proposed to address Q2. In this framework, geographical units and their traffic states are modeled as a graph with dynamic signals, and graph decomposition is achieved through two key modules: structure decomposition and signal decomposition. The structure decomposition module partitions the original graph into approximately orthogonal sub-graph structures by introducing soft graph mask matrices and structure independence regularization. The signal decomposition module utilizes a Residual Decomposition Flow to decouple graph signals into multiple sub-graph signal components. To ensure the independence of these components, a signal independence regularization mechanism is designed to minimize the mutual information between components. By generating orthogonal sub-graph structures with independent signals, the TGDL framework adheres to the CI Principle. Building on this, the TGDL framework assigns a dedicated prediction model to each sub-graph and aggregates the predictions from all sub-graphs to generate the final traffic forecast. As a general framework, TGDL is compatible with any graph-based traffic prediction model and significantly enhances its performance. Comprehensive experiments on four public traffic datasets validate the effectiveness of this theoretical approach and the TGDL framework.

6.2 Theoretical Analysis of Decomposition Prediction

This section primarily covers two aspects: first, analyzing prediction error in decomposition forecasting based on information theory; second, deriving sufficient conditions under which the prediction error of the decomposed problem is lower than that of the original problem.

6.2.1 *Traffic Prediction from an Information-Theoretic Perspective*

In the traffic state prediction problem, let X denote historical traffic states and Y denote future traffic states. The objective is to construct a prediction function $f(\cdot)$ capable of predicting Y given X, i.e.

$$\hat{Y} = f(X, \theta), \tag{6.1}$$

where $\hat{Y}$ is the prediction result, and θ represents the parameters of the function $f(\cdot)$, which are optimized by minimizing the prediction error across all training samples.

To analyze the underlying sources of prediction errors in the traffic prediction problem, we adopt the noisy channel model from information theory as follows:

$$Y \xrightarrow{\text{Encoder}} \tilde{X} \xrightarrow{\text{Noisy Channel}} X \xrightarrow{\text{Decoder}} \hat{Y}. \tag{6.2}$$

It considers a sender holding a source message Y that the receiver cannot directly observe. From the receiver's perspective, there is intrinsic uncertainty in determining Y. When a communication channel is established and the sender transmits Y, this uncertainty can be reduced or eliminated. To minimize the amount of transmitted data, the sender applies a lossless compression algorithm to produce a compressed message $\tilde{X}$, which is then transmitted. The receiver decodes $\tilde{X}$ to reconstruct $\hat{Y}$.

From an information theory perspective, traffic prediction can be modeled as message transmission over a noisy channel. This analogy is summarized in Table 6.1 and visually illustrated in Fig. 6.1. In Eq. (6.2), the decoding step corresponds to the traffic prediction process, where the Decoder represents the prediction function defined in Eq. (6.1).

In this model, the decoding step corresponds to traffic prediction, and the decoder corresponds to the prediction function defined in Eq. (6.1).

In the message transmission process defined by Eq. (6.2), the prediction error $Y \neq \hat{Y}$ primarily stems from two sources:

Table 6.1 Analogy between the noisy channel model and traffic prediction

Notation	Noisy channel model	Traffic prediction
$\hat{Y} = f(X, \theta)$	Decoder	Prediction model
Y	Source message	Future traffic state
$\tilde{X}$	Compressed message	—
X	Compressed message with noise	Historical traffic state
$\hat{Y}$	Reconstructed message	Predicted future traffic state

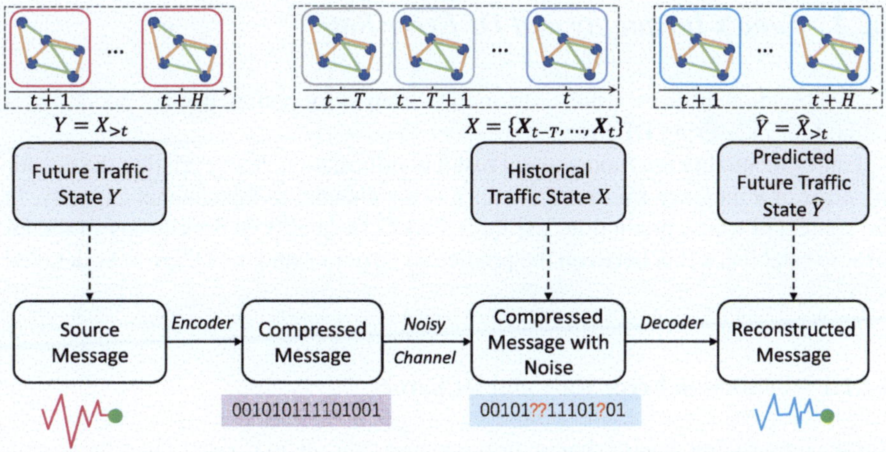

Fig. 6.1 Illustration of modeling traffic prediction as a transmission process over a noisy channel

- Noisy Channel: Channel noise causes information loss in X relative to its original state $\tilde{X}$, making it impossible for the decoder to fully reconstruct Y. This error reflects the missing information in historical traffic data X relative to future traffic data Y. It is thus termed Data-Induced (DI) error.
- Decoder Capability: Even if X retains complete information, the decoder may fail to accurately reconstruct Y due to limitations in modeling capability. Since the decoder corresponds to the prediction function $f(\cdot)$, this error is determined by the fitting capability of the prediction model and is termed Model-Induced (MI) error.

It is worth noting that in prediction models, the goal of parameter optimization is to minimize MI error. However, even if MI error is reduced to zero, DI error persists, representing the inherent performance upper bound of the prediction model on a given traffic dataset. This upper bound is a specific property of the dataset and is unaffected by the prediction model parameters. Therefore, it can be concluded that *DI error serves as an indicator of the prediction difficulty of a dataset*. In other words, a larger DI error indicates a more challenging dataset, while a smaller DI error suggests a relatively easier one.

In the following subsections, we will derive a sufficient condition to guarantee that the DI error in the decomposition prediction problem is lower than that in the original problem, thereby ensuring that the decomposition prediction problem is easier to solve.

6.2.2 *Fano's Inequality and DI Error Rate*

This subsection utilizes Fano's Inequality from information theory[1] to derive a method for measuring DI error.

Fano's Inequality is a fundamental result in information theory relating the decoding error of a message transmitted over a noisy channel to the channel capacity. In the context of traffic prediction, we apply Fano's Inequality to the entropy of the DI error to establish a link between the prediction error rate and the DI error, as detailed below.

6.2.2.1 Prediction Error Rate and DI Error

Before introducing Fano's Inequality, we first present two prerequisite concepts regarding prediction error: prediction error rate and the entropy of DI error.

Definition 6.1 (*Prediction Error Rate*) Assuming the traffic state variable Y is a discrete variable with M possible values,[2] based on the Markov chain in Eq. (6.2), the prediction error rate of a traffic prediction model can be defined as:

$$e = \Pr(\hat{Y} \neq Y), \tag{6.3}$$

where e is an end-to-end prediction error metric comprising both DI error and MI error.

Definition 6.2 (*Entropy of DI Error*) As described in Sect. 6.2.1, DI error represents the information loss of the future traffic state Y relative to the historical traffic state X. Therefore, DI error can be represented by the conditional entropy $\mathrm{H}(Y|X)$, i.e.,

$$\mathrm{H}(\text{DI Error}) = \mathrm{H}(Y|X), \tag{6.4}$$

where $\mathrm{H}(\cdot)$ denotes the entropy function.

Although $\mathrm{H}(Y|X)$ can measure the magnitude of DI error, the distributions of variables X and Y are typically unknown in real-world scenarios, making direct calculation of $\mathrm{H}(Y|X)$ difficult. Therefore, this chapter adopts Fano's Inequality [8] to establish the connection between the prediction error rate and the entropy of DI error, thereby estimating the DI error indirectly.

Theorem 6.1 (Fano's Inequality) Assume $Y \rightarrow X \rightarrow \hat{Y}$ is a Markov chain, where Y and $\hat{Y}$ are discrete random variables with M distinct values. Let $e = \Pr(\hat{Y} \neq Y)$. Then the relationship between e and $\mathrm{H}(Y|X)$ can be described as:

---

[1] Introduction to Fano's Inequality: https://en.wikipedia.org/wiki/Fano%27s_inequality.

[2] For datasets with continuous values, this assumption can be satisfied through discretization.

$$\mathrm{H}(Y|X) \leq \mathrm{H}(e) + e\log(M-1), \tag{6.5}$$

where $\mathrm{H}(e)$ is the entropy of the error rate, calculated as $\mathrm{H}(e) = -e \cdot \log e - (1 - e) \cdot \log(1-e)$.

Proof First, define a binary random variable E representing the event of a prediction error:

$$E = \begin{cases} 1, & \text{if } Y \neq \hat{Y}, \\ 0, & \text{if } Y = \hat{Y}. \end{cases} \tag{6.6}$$

Using the chain rule for entropy [4], the conditional entropy $\mathrm{H}(E, Y|\hat{Y})$ can be expanded as:

$$\mathrm{H}(E, Y|\hat{Y}) = \mathrm{H}(Y|\hat{Y}) + \underbrace{\mathrm{H}(E|Y, \hat{Y})}_{=0} = \mathrm{H}(E|\hat{Y}) + \mathrm{H}(Y|E, \hat{Y}). \tag{6.7}$$

For the right side of the equation, since the second term on the left $\mathrm{H}(E|Y, \hat{Y}) = 0$, the equation simplifies to:

$$\mathrm{H}(Y|\hat{Y}) = \mathrm{H}(E|\hat{Y}) + \mathrm{H}(Y|E, \hat{Y}). \tag{6.8}$$

Next, let p_e denote the probability of prediction error. Expanding the rightmost term $\mathrm{H}(Y|E, \hat{Y})$ in Eq. (6.8) yields:

$$\mathrm{H}(Y|E, \hat{Y}) = p_e \cdot \mathrm{H}(Y|E=1, \hat{Y}) + (1-p_e) \cdot \underbrace{\mathrm{H}(Y|E=0, \hat{Y})}_{=0} = p_e \cdot \mathrm{H}(Y|E=1, \hat{Y}). \tag{6.9}$$

Here, since $E = 1$ implies $Y \neq \hat{Y}$, given a message Y with M distinct values, the possible values for Y are narrowed down to $M - 1$ distinct values. Thus, an upper bound for the conditional entropy $\mathrm{H}(Y|E, \hat{Y})$ is obtained:

$$\mathrm{H}(Y|E, \hat{Y}) \leq \log(M-1) \cdot p_e. \tag{6.10}$$

For the term $\mathrm{H}(E|\hat{Y})$ in Eq. (6.8), based on entropy properties:

$$\mathrm{H}(E|\hat{Y}) \leq \mathrm{H}(E) = \mathrm{H}(p_e). \tag{6.11}$$

Substituting Eqs. (6.10) and (6.11) into Eq. (6.8) gives:

$$\mathrm{H}(Y|\hat{Y}) \leq \mathrm{H}(p_e) + \log(M-1) \cdot p_e. \tag{6.12}$$

According to the Data Processing Inequality [1], for the Markov chain $Y \to X \to \hat{Y}$, it follows that $\mathrm{H}(Y|\hat{Y}) \geq \mathrm{H}(Y|X)$. Combining this with Eq. (6.12) proves Fano's Inequality in Eq. (6.5). □

Through the above proof, it is evident that the error rate e and the DI error entropy $\mathrm{H}(Y|X)$ are linked via Fano's Inequality.

6.2.2.2 Lower Bound of Error Rate

Based on Fano's Inequality, a lower bound for the error rate can be further derived. First, define a function S related to the prediction error rate e as:

$$S(e) = \mathrm{H}(e) + e \log(M - 1). \tag{6.13}$$

Considering the monotonicity of this function, we propose the following lemma.

Lemma 6.1 *When the condition $e < 1 - 1/M$ holds, the function $S(e)$ is monotonically increasing.*

Proof To prove this lemma, we first calculate the second derivative of the function $S(e)$:

$$\frac{\partial^2 S(e)}{\partial e^2} = -\frac{1}{e(1 - e)}. \tag{6.14}$$

Since $e \in [0, 1]$, $\frac{\partial^2 S(e)}{\partial e^2} < 0$ always holds. Thus, $S(e)$ is strictly concave and achieves its maximum where its first derivative is zero. The first derivative of $S(e)$ is:

$$\frac{\partial S(e)}{\partial e} = \log(1 - e) - \log(e) - \log(M - 1). \tag{6.15}$$

Setting the first derivative to zero yields:

$$e = 1 - \frac{1}{M}. \tag{6.16}$$

Combined with the concave property of $S(e)$, when the error rate $e < 1 - 1/M$, $S(e)$ is a monotonically increasing function. □

Since variable Y has M distinct values, the error rate e cannot exceed $1 - 1/M$. Thus, Eq. (6.13) can be directly regarded as a monotonically increasing function. For convenience, let $S^{\min} = \mathrm{H}(Y|X)$. According to Fano's Inequality in Eq. (6.5), we have $\mathrm{H}(Y|X) \leq S(e)$, making $S^{\min}$ the lower bound of function $S(e)$. As $S(e)$ is monotonically increasing, $S^{\min}$ uniquely corresponds to a lower bound of the error rate, denoted as $e^{\min}$, defined by:

$$S^{\min} = S(e^{\min}) = \mathrm{H}(e^{\min}) + e^{\min} \log(M - 1). \tag{6.17}$$

6.2.2.3 Data-Induced Error Rate

According to Eq. (6.17), there exists a monotonically increasing implicit function $e^{\min} = f^{-1}(S^{\min})$, where $e^{\min}$ is the lower bound of the error rate e, i.e., $e \geq e^{\min}$. Furthermore, $e^{\min}$ is determined entirely by the entropy of the DI error, indicating that $e^{\min}$ represents the portion of the overall prediction error rate e caused by data. In other words, it provides a standard to measure the prediction error rate resulting from DI error. Therefore, this chapter refers to $e^{\min}$ as the DI Error Rate.

6.2.3 DI Error Rate in Decomposition Algorithms

This section focuses on analyzing the relationship between the DI error rate of the original prediction problem and that of the decomposed sub-problems.

Before delving into the analysis, a formal definition of the decomposition algorithm for traffic prediction problems is provided.

Definition 6.3 (*Decomposition Algorithm*) For a traffic prediction problem, given input and output data variables X and Y, a decomposition algorithm is defined as a function $f(\cdot)$ such that $\{X_1, \ldots, X_k, \ldots, X_K\} = f(X)$ and $\{Y_1, \ldots, Y_k, \ldots, Y_K\} = f(Y)$, where X_k and Y_k denote the k-th sub-component obtained from decomposition. During this process, the decomposition algorithm ensures that $\{X_1, \ldots, X_K\}$ and $\{Y_1, \ldots, Y_K\}$ can completely reconstruct X and Y respectively. Simultaneously, this data decomposition and reconstruction process forms two Markov chains: $X \rightarrow \{X_1, \ldots, X_K\} \rightarrow X$ and $Y \rightarrow \{Y_1, \ldots, Y_K\} \rightarrow Y$.

Based on the above definition of the decomposition algorithm, the following lemma can be derived.

Lemma 6.2 *Given the Markov chain of the decomposition algorithm* $X \rightarrow \{X_1, \ldots, X_K\} \rightarrow X$, *it holds that* $\mathrm{H}(X) = \mathrm{H}(X_1, \ldots, X_K)$. *Similarly,* $\mathrm{H}(Y) = \mathrm{H}(Y_1, \ldots, Y_K)$ *is also obtained.*

Proof For convenience, let $\{X_1, \ldots, X_K\}$ be denoted as Z. According to the Data Processing Inequality [1]:

$$\mathrm{I}(X; Z) \geq \mathrm{I}(X; X), \tag{6.18}$$

where $\mathrm{I}(\cdot; \cdot)$ represents the mutual information between two variables. Since the relationship $\mathrm{I}(X; X) = \mathrm{H}(X)$ exists, we have:

$$\mathrm{I}(X; Z) \geq \mathrm{H}(X). \tag{6.19}$$

According to the definition of mutual information:

$$\mathrm{I}(X; Z) = \mathrm{H}(X) - \mathrm{H}(X|Z) \leq \mathrm{H}(X), \tag{6.20}$$

where $\mathrm{H}(X|Z)$ is conditional entropy and $\mathrm{H}(X|Z) \geq 0$. Combining Eq. (6.19) and Eq. (6.20), we derive:

$$\mathrm{I}(X; Z) = \mathrm{H}(X), \tag{6.21}$$

which implies $\mathrm{H}(X|Z) = 0$. This indicates:

$$\mathrm{H}(X) = \mathrm{H}(Z) = \mathrm{H}(X_1, \ldots, X_K). \tag{6.22}$$

Thus, the equality $\mathrm{H}(X) = \mathrm{H}(X_1, \ldots, X_K)$ is proven. Similarly, denoting $\{Y_1, \ldots, Y_K\}$ as Z, we can prove $\mathrm{H}(Y) = \mathrm{H}(Y_1, \ldots, Y_K)$ through a similar process. □

According to Lemma 6.2, $S^{\min} = \mathrm{H}(Y|X)$ can be further associated with the decomposed sub-components:

$$S^{\min} = \mathrm{H}(Y|X) = \mathrm{H}\big(Y_1, \ldots, Y_K|X\big) = \mathrm{H}\big(Y_1, \ldots, Y_K|X_1, \ldots, X_K\big). \tag{6.23}$$

For the k-th sub-problem obtained via decomposition, it also possesses a noisy channel Markov chain $Y_k \to X_k \to \hat{Y}_k$. Let $S_k^{\min} = \mathrm{H}(Y_k|X_k)$. According to Eq. (6.17), sub-problem k also has a data-induced error rate $e_k^{\min}$, which is an implicit function of $S_k^{\min}$. We propose the following theorem regarding $e_k^{\min}$ of the sub-problem and $e^{\min}$ of the original problem.

Theorem 6.2 (Condition for $e^{\min} \geq e_k^{\min}$) Assume $Y \in \{1, \ldots, M\}$ and $Y_k \in \{1, \ldots, M\}$ for all $k \in \{1, \ldots, K\}$. Additionally, X and X_k are generated by the processes $Y \to X$ and $Y_k \to X_k$ respectively, and variables X, Y, X_k, Y_k satisfy Eq. (6.23). Then, if $Y_k \perp X_{k'}$ holds for all $k' \neq k$, the inequality $e^{\min} \geq e_k^{\min}$ holds.

Proof Based on Eq. (6.23) and the property of joint entropy[3], we obtain:

$$S^{\min} = \mathrm{H}\big(Y_1, \ldots, Y_K|X\big) \geq \max_{k \in \{1,\ldots,K\}} \{\mathrm{H}(Y_k|X)\} \geq \mathrm{H}(Y_k|X). \tag{6.24}$$

The inequality in Eq. (6.24) holds because the joint entropy of a set of variables is greater than or equal to the entropy of any single variable [4]. For conditional entropy $\mathrm{H}(Y_k|X)$, if $Y_k \perp X_{k'}$ holds for all $k' \neq k$, then:

$$\mathrm{H}(Y_k|X) = \mathrm{H}(Y_k|X_1, \ldots, X_K) = \mathrm{H}(Y_k|X_k) = S_k^{\min}. \tag{6.25}$$

Combining Eqs. (6.24) and (6.25), we get $S^{\min} \geq S_k^{\min}$. According to Lemma 6.1, since $S(e)$ is a monotonically increasing function, the inequality $e^{\min} \geq e_k^{\min}$ holds. The proof is complete. □

Building on Theorem 6.2, this chapter proposes a stricter theorem guaranteeing that $e_k^{\min}$ is strictly less than $e^{\min}$.

[3] Properties of joint entropy: https://en.wikipedia.org/wiki/Joint_entropy.

Theorem 6.3 (Condition for $e^{\min} > e_k^{\min}$) Based on the conditions of Theorem 6.2, if $Y_k \perp Y_{k'}$ holds for all $k' \neq k$, then the inequality $e^{\min} > e_k^{\min}$ also holds.

Proof Without loss of generality, assume sorting $Y_1, \ldots, Y_K$ by entropy magnitude yields $\{Y_{m_1}, Y_{m_2}, \ldots, Y_{m_K}\}$, where $\mathrm{H}(Y_{m_1}) \geq \mathrm{H}(Y_{m_2}) \geq \cdots \geq \mathrm{H}(Y_{m_K})$. Then, the joint entropy in Eq. (6.24) can be expressed as:

$$\mathrm{H}(Y_1, \ldots, Y_K) = \mathrm{H}\left(Y_{m_1}, \ldots, Y_{m_K}\right) = \mathrm{H}(Y_{m_1}) + \mathrm{H}(Y_{m_2}|Y_{m_1}) + \cdots + \mathrm{H}(Y_{m_K}|Y_{m_1}, \ldots, Y_{m_{K-1}}), \tag{6.26}$$

Note that the condition X in conditional probabilities is omitted here for brevity.

If $Y_k \perp Y_{k'}$ holds for all $k' \neq k$, meaning $Y_{m_1} \perp Y_{m_2}$, then $\mathrm{H}(Y_{m_2}|Y_{m_1}) = \mathrm{H}(Y_{m_2})$. Thus, the joint entropy relationship can be derived as:

$$\mathrm{H}(Y_1, \ldots, Y_K) = \mathrm{H}(Y_{m_1}) + \mathrm{H}(Y_{m_2}) + \cdots + \mathrm{H}(Y_{m_K}). \tag{6.27}$$

Since variables $\{Y_1, \ldots, Y_K\}$ are mutually independent and the entropy of each variable $\mathrm{H}(Y_{m_k}) \geq 0$, there are at least $K - 1$ variables with entropy greater than zero. This implies:

$$\mathrm{H}(Y_1, \ldots, Y_K) > \mathrm{H}(Y_{m_1}) = \max_{k \in \{1, \ldots, K\}} \{\mathrm{H}(Y_k)\} \geq \mathrm{H}(Y_k). \tag{6.28}$$

Substituting Eq. (6.28) into Eq. (6.24) in the proof of Theorem 6.2, we derive $S^{\min} > S_k^{\min}$. Consequently, it follows that the error rate of the original problem $e^{\min}$ is strictly greater than the error rate of the sub-problem $e_k^{\min}$, i.e., $e^{\min} > e_k^{\min}$. □

Conclusion 1: Theorem 6.3 provides a sufficient condition for $e^{\min} > e_k^{\min}$. Specifically, if $Y_k \perp X_{k'}$ and $Y_k \perp Y_{k'}$ hold for all $k' \neq k$, this condition is satisfied. This implies that *if a data decomposition algorithm ensures that data in any decomposed sub-problem is independent of data in other sub-problems, the prediction difficulty of each sub-problem (measured by the DI error rate) will be lower than that of the original problem.*

6.2.4 DI Error Value in Decomposition Algorithms

Section 6.2.3 established the relationship between the DI error rate of the original problem and that of each sub-problem. However, two issues remain unresolved. First, as described in Definition 6.3, the final prediction of the original problem is achieved by aggregating results from all sub-problems; thus, the DI error of the aggregated prediction must be compared with that of the original prediction. Second, while prediction error rate treats traffic prediction as a discrete problem, it is typically treated as a continuous value prediction problem in practice, making direct application of error rate difficult. Therefore, this section focuses on addressing these two issues.

6.2.4.1 Prediction Error Value

For a continuous prediction problem where the true value is $Y \in \mathbb{R}$ and the predicted value is $\hat{Y} \in \mathbb{R}$, the error value can be defined as:

$$\xi = \mathbb{E}\left[\left(Y - \hat{Y}\right)^2\right]. \tag{6.29}$$

In the predictive learning framework [23], prediction error can be divided into *error certainty* and *error magnitude*. Error certainty refers to the probability of prediction error occurring, defined as the error rate $e = \Pr(Y \neq \hat{Y})$, while error magnitude measures the size of the error.

Let $Y^{(e)}$ and $\hat{Y}^{(e)}$ denote the variables corresponding to the true and predicted values when $Y \neq \hat{Y}$, respectively. The error magnitude can be defined as:

$$\mathbb{E}[\delta] = \mathbb{E}\left[\left(Y^{(e)} - \hat{Y}^{(e)}\right)^2\right]. \tag{6.30}$$

Using error certainty e and error magnitude $\mathbb{E}[\delta]$, the prediction error value ξ can be expressed as:

$$\xi = e \cdot \mathbb{E}[\delta]. \tag{6.31}$$

6.2.4.2 DI Error Magnitude

According to the error source analysis in Sect. 6.2.1, error magnitude should consist of MI components and DI components. Thus, the error magnitude in Eq. (6.30) can be expressed as:

$$\mathbb{E}[\delta] = \mathbb{E}\left[\left(Y^{(e)} - \left(Y^{(e)} + \Delta^{\min} + \Delta^{\text{model}}\right)\right)^2\right], \tag{6.32}$$

where $\Delta^{\min}$ and Δ^{model} represent the DI and MI components of the error magnitude, respectively. Therefore, the data-induced error component of the original traffic prediction problem—the *DI error magnitude*—can be expressed as:

$$\mathbb{E}[\delta^{\min}] = \mathbb{E}\left[\left(Y^{(e)} - \left(Y^{(e)} + \Delta^{\min}\right)\right)^2\right] = \mathbb{E}\left[\left(\Delta^{\min}\right)^2\right]. \tag{6.33}$$

Assuming the decomposition algorithm does not reduce the DI error magnitude, the DI error magnitude component of all decomposed predictions satisfies $\Delta^{\min} = \sum_{k=1}^{K} \Delta_k^{\min}$, where $\Delta_k^{\min}$ is the DI error magnitude component of the k-th component. Analogous to Eq. (6.33), the DI error magnitude of the k-th sub-problem is defined as $\mathbb{E}[\delta_k^{\min}] = \mathbb{E}[(\Delta_k^{\min})^2]$.

Based on the above assumption, this chapter proposes the following lemma.

Lemma 6.3 *Given mutually independent decomposed sub-components* $\{Y_1, \ldots, Y_K\}$*, if variables* $\{X_1, \ldots, X_K\}$ *are also mutually independent, then* $\mathbb{E}[\delta^{\min}] = \sum_{k=1}^{K} \mathbb{E}[\delta_k^{\min}]$*, where* $\mathbb{E}[\delta_k^{\min}]$ *is the DI error magnitude of the k-th sub-problem.*

Proof For the noisy channel model described in Eq. (6.2), the message X contains the original message Y and the data-induced error $\Delta^{\min}$, implying their conditional entropy is zero, i.e.,

$$\mathrm{H}(Y, \Delta^{\min}|X) = \mathrm{H}(X|Y, \Delta^{\min}) = 0. \tag{6.34}$$

Under the "Decompose-Then-Predict" premise, if the k-th sub-problem is viewed as a noisy channel model described by the Markov chain $Y_k \rightarrow \tilde{X}_k \rightarrow X_k \rightarrow \hat{Y}_k$, then for this sub-problem, we can derive $\mathrm{H}(Y_k, \Delta_k^{\min}|X_k) = \mathrm{H}(X_k|Y_k, \Delta_k^{\min}) = 0$.

Next, we prove that for all $k \neq k'$, $\Delta_k^{\min} \perp \Delta_{k'}^{\min}$ holds. Define $B_k = (Y_k, \Delta_k^{\min})$. Based on the properties of joint entropy, we can derive:

$$\mathrm{H}(X_k, B_k) = \mathrm{H}(B_k) + \mathrm{H}(X_k|B_k) = \mathrm{H}(X_k) + \mathrm{H}(B_k|X_k). \tag{6.35}$$

Since $\mathrm{H}(X_k|B_k) = 0$ and $\mathrm{H}(B_k|X_k) = 0$, we obtain:

$$\mathrm{H}(B_k) = \mathrm{H}(X_k). \tag{6.36}$$

Then, given the conditional entropy $\mathrm{H}(X_k, B_k|X_{k'})$ for a sub-problem, a similar process yields:

$$\mathrm{H}(X_k, B_k|X_{k'}) = \mathrm{H}(B_k|X_{k'}) + \mathrm{H}(X_k|B_k, X_{k'}) = \mathrm{H}(X_k|X_{k'}) + \mathrm{H}(B_k|X_k, X_{k'}). \tag{6.37}$$

Since $\mathrm{H}(X_k|B_k, X_{k'}) = 0$ and $\mathrm{H}(B_k|X_k, X_{k'}) = 0$, we deduce:

$$\mathrm{H}(B_k|X_{k'}) = \mathrm{H}(X_k|X_{k'}). \tag{6.38}$$

Similarly, a mirrored relationship can be derived:

$$\mathrm{H}(X_k|B_{k'}) = \mathrm{H}(B_k|B_{k'}). \tag{6.39}$$

Substituting Eqs. (6.36) and (6.38) into the mutual information equation $\mathrm{I}(B_k; X_{k'}) = \mathrm{H}(B_k) - \mathrm{H}(B_k|X_{k'})$ gives:

$$\mathrm{I}(B_k; X_{k'}) = \mathrm{H}(X_k) - \mathrm{H}(X_k|X_{k'}) = \mathrm{I}(X_k; X_{k'}). \tag{6.40}$$

Since $X_k \perp X_{k'}$, we can conclude:

$$\mathrm{I}(B_k; X_{k'}) = 0. \tag{6.41}$$

Similarly, the mirrored relationship $\mathrm{I}(B_{k'}; X_k) = 0$ is obtained, thus $\mathrm{I}(B_k; X_{k'}) = \mathrm{I}(B_{k'}; X_k)$.

From the equality $\mathrm{I}(B_k; X_{k'}) = \mathrm{I}(B_{k'}; X_k)$, we infer:

$$\mathrm{H}(B_k) - \mathrm{H}(B_k|X_{k'}) = \mathrm{H}(X_k) - \mathrm{H}(X_k|B_{k'}). \tag{6.42}$$

Based on the conclusion from Eq. (6.36) that $\mathrm{H}(B_k) = \mathrm{H}(X_k)$, simple derivation shows:

$$\mathrm{H}(B_k|X_{k'}) = \mathrm{H}(X_k|B_{k'}). \tag{6.43}$$

Substituting Eqs. (6.36), (6.38), (6.39), and (6.43) into $\mathrm{I}(B_k; B_{k'})$ yields:

$$\begin{aligned} \mathrm{I}(B_k; B_{k'}) &= \mathrm{H}(B_k) - \mathrm{H}(B_k|B_{k'}) = \mathrm{H}(X_k) - \mathrm{H}(X_k|B_{k'}) \\ &= \mathrm{H}(X_k) - \mathrm{H}(B_k|X_{k'}) = \mathrm{H}(X_k) - \mathrm{H}(X_k|X_{k'}) \\ &= \mathrm{I}(X_k; X_{k'}) = 0. \end{aligned} \tag{6.44}$$

From the above derivation, we conclude:

$$\forall k \neq k', \, B_k \perp B_{k'}. \tag{6.45}$$

Since B_k is defined as $B_k = (Y_k, \Delta_k^{\min})$, and given $Y_k \perp Y_{k'}$, the independence relationship $B_k \perp B_{k'}$ implies:

$$\forall k \neq k', \, \Delta_k^{\min} \perp \Delta_{k'}^{\min}. \tag{6.46}$$

Thus far, it is proven that *data-induced error components* $\Delta_k^{\min}$ *are mutually independent across all sub-problems.*

Next, we integrate the DI error components of all sub-problems and compare them with that of the original problem. First, using the error decomposition equation $\Delta^{\min} = \sum_{k=1}^{K} \Delta_k^{\min}$, expand the expectation of the squared error:

$$\begin{aligned} \mathbb{E}[\delta^{\min}] &= \mathbb{E}\left[\left(\Delta^{\min}\right)^2\right] = \mathbb{E}\left[\left(\Delta_1^{\min} + \cdots + \Delta_K^{\min}\right)^2\right] \\ &= \mathbb{E}\left[\sum_{k=1}^{K} \left(\Delta_k^{\min}\right)^2 + 2 \sum_{1 \le i < j \le K} \Delta_i^{\min} \cdot \Delta_j^{\min}\right]. \end{aligned} \tag{6.47}$$

Applying the linearity of expectation and combining like terms yields:

$$\mathbb{E}[\delta^{\min}] = \sum_{k=1}^{K} \mathbb{E}\left[\left(\Delta_k^{\min}\right)^2\right] + 2 \underbrace{\sum_{1 \le i < j \le K} \mathbb{E}\left[\Delta_i^{\min} \cdot \Delta_j^{\min}\right]}_{\text{Cross Term}=0}. \tag{6.48}$$

Since $\Delta_k^{\min} \perp \Delta_{k'}^{\min}$ holds for all $k \neq k'$, the cross term in Eq. (6.48) equals zero. Therefore, removing this term simplifies Eq. (6.48) to:

$$\mathbb{E}[\delta^{\min}] = \sum_{k=1}^{K} \mathbb{E}\left[\left(\Delta_k^{\min}\right)^2\right] = \sum_{k=1}^{K} \mathbb{E}[\delta_k^{\min}]. \tag{6.49}$$

This proves the equality $\mathbb{E}[\delta^{\min}] = \sum_{k=1}^{K} \mathbb{E}[\delta_k^{\min}]$. □

6.2.4.3 DI Error Value

For a given prediction problem, the *DI error value* of the original problem and its decomposed sub-problems is defined as:

$$\xi^{\min} = e^{\min} \cdot \mathbb{E}[\delta^{\min}], \qquad \xi_k^{\min} = e_k^{\min} \cdot \mathbb{E}[\delta_k^{\min}], \tag{6.50}$$

where $\xi^{\min}$ denotes the error value of the original problem, and $\xi_k^{\min}$ denotes the error value of the k-th sub-problem.

The relationship between the DI error value of the original problem $\xi^{\min}$ and the aggregated DI error value of all decomposed sub-problems $\sum_{k=1}^{K} \xi_k^{\min}$ is given by the following theorem.

Theorem 6.4 (Condition for $\xi^{\min} > \sum_{k=1}^{K} \xi_k^{\min}$) For $k \in \{1, \ldots, K\}$, let $Y \in \{1, \ldots, M\}$, $Y_k \in \{1, \ldots, M\}$, $\hat{Y}_k \in \{1, \ldots, M\}$, and let X and X_k be variables generated by Markov chains $Y \to X$ and $Y_k \to X_k$ respectively. Assume X, Y, X_k, Y_k satisfy Eq. (6.23). Then, if the following conditions are met, the inequality $\xi^{\min} > \sum_{k=1}^{K} \xi_k^{\min}$ holds:

(i) For any Y_k, $Y_k \perp X_{k'}$, where $k' \neq k$;
(ii) For any pair of variables Y_k and $Y_{k'}$ in $\{Y_1, \ldots, Y_K\}$, $Y_k \perp Y_{k'}$;
(iii) For any pair of variables X_k and $X_{k'}$ in $\{X_1, \ldots, X_K\}$, $X_k \perp X_{k'}$.

Proof Based on Eq. (6.50), the aggregated DI error value for all sub-problems can be expressed as:

$$\sum_{k=1}^{K} \xi_k^{\min} = \sum_{k=1}^{K} e_k^{\min} \cdot \mathbb{E}[\delta_k^{\min}] < \sum_{k=1}^{K} e^{\min} \cdot \mathbb{E}[\delta_k^{\min}] = e^{\min} \sum_{k=1}^{K} \mathbb{E}[\delta_k^{\min}], \tag{6.51}$$

where the inequality holds by applying Theorem 6.3. This theorem states that if conditions (i) and (ii) are met, then $e_k^{\min} < e^{\min}$.

Next, according to condition (iii), variables $\{X_1, \ldots, X_K\}$ are mutually independent. Combined with Lemma 6.3, we deduce:

$$\mathbb{E}[\delta^{\min}] = \sum_{k=1}^{K} \mathbb{E}[\delta_k^{\min}]. \tag{6.52}$$

Substituting this equation into Eq. (6.51), we obtain:

$$e^{\min} \sum_{k=1}^{K} \mathbb{E}[\delta_k^{\min}] = e^{\min} \cdot \mathbb{E}[\delta^{\min}] = \xi^{\min}. \tag{6.53}$$

From Eqs. (6.51) and (6.53), it follows that:

$$\xi^{\min} > \sum_{k=1}^{K} \xi_k^{\min}. \tag{6.54}$$

The theorem is thus proven. □

Theorem 6.4 establishes the relationship between the DI error in the original problem and the decomposed sub-problems (cross-sub-problem aggregated prediction) in continuous prediction settings, effectively resolving the two key issues raised at the beginning of this section.

Conclusion 2: According to Theorem 6.4, *if there exists a decomposition algorithm ensuring that for all $k' \neq k$, $Y_k \perp X_{k'}$, $Y_k \perp Y_{k'}$, and $X_k \perp X_{k'}$, then the total DI error of the decomposed prediction is less than that of the original prediction.* This indicates that effective decomposition algorithms can reduce the difficulty of prediction problems. In such cases, if a sufficiently effective prediction model can eliminate all model-induced MI errors, the performance of the decomposed prediction is guaranteed to exceed that of the original prediction.

In traffic prediction models, variables X and Y represent traffic state time series, i.e., $\{X^{(1)}, \dots, X^{(t)}, Y^{(t+1)}\}$, where $X^{(t)}$ denotes the sample of X at time t. To satisfy the conditions in Theorem 6.4, the decomposition algorithm must ensure that sequences $\{X_k^{(1)}, \dots, X_k^{(t)}, Y_k^{(t+1)}\}$ and $\{X_{k'}^{(1)}, \dots, X_{k'}^{(t)}, Y_{k'}^{(t+1)}\}$ are mutually independent for all $k \neq k'$.

Based on the above analysis, this chapter establishes a guiding principle for designing decomposition algorithms.

Principle 6.1 (Component Independence (CI) Principle) *In the "Decompose-Then-Predict" paradigm, effective decomposition algorithms should maximize the independence of decomposed components. This is referred to as the CI Principle.*

This principle will guide the design of model regularization terms in the next section to constrain the decomposition process to satisfy the CI Principle.

6.2.5 Existence of Independent Decomposition

The CI Principle posits that a decomposition algorithm should maximize the independence among components to enhance data predictability. Yet, real-world traffic data inherently contain both explicit and latent correlations, raising a natural question: *Is it possible to decompose a traffic state series X into multiple statistically independent components?*

To address this, we assume the traffic state series X is generated from N underlying real-world latent patterns $Z_1, \ldots, Z_N$. The relationship can be expressed as:

$$X = \sum_{n=1}^{N} Z_n. \tag{6.55}$$

Furthermore, we assume each pattern Z_n consists of two components: an independent factor $Z_n^{(\text{ind})}$ unique to the pattern, and a shared correlated factor $Z^{(\text{corr})}$ representing global dependencies. Formally:

$$Z_n = Z_n^{(\text{ind})} + \alpha_n Z^{(\text{corr})}, \quad \text{where} \quad Z_{n_1}^{(\text{ind})} \perp Z_{n_2}^{(\text{ind})}, \ \forall n_1 \neq n_2. \tag{6.56}$$

Here, the coefficients $\alpha_n \in [0, 1]$ satisfy $\sum_{n=1}^{N} \alpha_n = 1$, and the shared factor $Z^{(\text{corr})}$ is independent of all independent factors. Based on this model, we propose the following theorem:

Theorem 6.5 (Existence of Independent Decomposition) Based on Eq. (6.56) in the main body, the traffic state series X can be decomposed into a set of mutually independent components:

$$X = X^{(1)} + \ldots + X^{(K)}, \quad \text{where} \quad X^{(k_1)} \perp X^{(k_2)}, \ \forall k_1 \neq k_2, \tag{6.57}$$

where $X^{(k)}$ denotes the k-th component obtained from the decomposition of the original traffic state series X.

Proof Without loss of generality, assume the traffic state X is generated from two underlying real-world patterns Z_1 and Z_2, i.e.,

$$X = Z_1 + Z_2. \tag{6.58}$$

Each pattern consists of an independent factor and a correlated factor:

$$Z_1 = Z_1^{(\text{ind})} + \alpha Z^{(\text{corr})}, \quad Z_2 = Z_2^{(\text{ind})} + \beta Z^{(\text{corr})}, \tag{6.59}$$

where $Z_1^{(\text{ind})} \perp Z_2^{(\text{ind})}$. Substituting Eq. (6.59) into Eq. (6.58), we obtain

$$X = Z_1^{(\text{ind})} + Z_2^{(\text{ind})} + (\alpha + \beta) Z^{(\text{corr})}. \tag{6.60}$$

Hence, X can be decomposed into three mutually independent components:

$$X^{(1)} = Z_1^{(\text{ind})}, \quad X^{(2)} = Z_2^{(\text{ind})}, \quad X^{(3)} = (\alpha + \beta) Z^{(\text{corr})}. \tag{6.61}$$

Since all components are statistically independent (as $Z^{(\text{corr})}$ is independent of individual independent factors), the decomposition of X into mutually independent subcomponents exists. □

This theorem implies that even in the presence of shared correlated factors, it is theoretically feasible to decompose the traffic state X into multiple statistically independent components. This finding provides a solid theoretical foundation for the applicability of the CI Principle.

6.3 Graph Decomposition Learning Framework

This section proposes a Theory-guided Graph Decomposition Learning (TGDL) framework aimed at enhancing the performance of existing models in traffic scenarios under the "Decompose-Then-Predict" paradigm. The CI Principle states that decomposing a traffic prediction problem into mutually independent subcomponents can effectively reduce data-induced errors and thus improve predictability. This provides a theoretical foundation for designing decomposition-based learning frameworks. Guided by this principle, our TGDL framework is designed to produce traffic subgraphs that are as independent as possible in both structure and signal. This theoretical goal is achieved through two decomposition modules and two regularization terms:

- **Structure Decomposition Module (Sect.** 6.3.2) partitions the original traffic graph into multiple structurally independent subgraphs. The corresponding *Structure Independence Regularization* (Eq. (6.67)) penalizes structural overlaps among subgraphs by minimizing the pairwise Frobenius inner product of their adjacency matrices.
- **Signal Decomposition Module (Sect.** 6.3.3) disentangles the dynamic graph signals into multiple temporal components aligned with the subgraphs. The *Signal Independence Regularization* (Eq. (6.76)) minimizes mutual information between the extracted representations of signal components.

6.3.1 Basic Concepts and Problem Definition

The TGDL framework organizes traffic state data as graph data with dynamic signals, termed a Traffic Graph, primarily consisting of graph structure and graph signals. First, the definition of the traffic graph structure is provided.

Definition 6.4 (*Traffic Graph Structure*) The traffic graph structure can be represented as $G = (\mathcal{V}, A)$, where $\mathcal{V} = \{v_1, \ldots, v_I\}$ is the set of nodes, with each node representing a geographical unit (e.g., a road segment or district). The adjacency matrix $\boldsymbol{A} \in \mathbb{R}^{|\mathcal{V}|\times|\mathcal{V}|}$ contains edge weights, where the element at the (i, j)-th position, a_{ij}, represents the weight of the edge from node v_i to node v_j.

Edge weights here can be calculated using various methods, such as geographical distance [15, 30], correlation between node features [9, 14], or other relevant metrics. Besides structure, the graph signal of the traffic graph is defined by dynamic traffic states.

Definition 6.5 (*Traffic Graph Signal*) Given a traffic graph G, the signal of node v_i at time step t is a vector $\mathbf{x}_{i,t} \in \mathbb{R}^F$, where elements of $\mathbf{x}_{i,t}$ represent traffic states, and F is the number of traffic state channels (e.g., inflow, outflow, speed). The signal of the entire graph at time t is a matrix $\mathbf{X}_t \in \mathbb{R}^{|\mathcal{V}|\times F}$, where the i-th row corresponds to $\mathbf{x}_{i,t}$.

Based on the definitions of traffic graph structure and graph signals, the traffic prediction problem is formally defined below.

Definition 6.6 (*Original Prediction Problem*) Given a traffic graph G, its historical signals within the time window $\{t - T, \ldots, t\}$ are denoted as $\mathcal{X} = \{\mathbf{X}_{t-T}, \ldots, \mathbf{X}_t\}$. Future traffic states for time steps greater than t are denoted as $\mathbf{Y}_{>t}$. The original traffic state prediction problem can be formalized as:

$$\hat{\mathbf{Y}}_{>t} = f_g\left(\mathcal{X}_t, G, \Theta\right), \tag{6.62}$$

where $\hat{\mathbf{Y}}_{>t}$ is the prediction of future traffic states, and Θ represents learnable parameters.

The original prediction problem learns to predict future traffic states directly from the input traffic graph and historical signals. In contrast, in the "Decompose-Then-Predict" paradigm, input data needs to be split into multiple sub-components via a decomposition step, serving as inputs for each sub-problem. Thus, the definition of the decomposition prediction problem differs.

Definition 6.7 (*Decomposition Prediction Problem*) Given an original traffic graph $G^{(g)} = (\mathcal{V}, \boldsymbol{A}^{(g)})$ and its historical signals $\mathcal{X}_t^{(g)}$, the decomposition algorithm partitions the original graph into K sub-graph components, denoted as $\{G^{(1)}, \ldots, G^{(K)}\}$, with corresponding graph signals $\{\mathcal{X}_t^{(1)}, \ldots, \mathcal{X}_t^{(K)}\}$. These sub-graphs satisfy:

$$\boldsymbol{A}^{(g)} = \sum_{k=1}^{K} \boldsymbol{A}^{(k)} \quad \text{and} \quad \mathcal{X}_t^{(g)} = \sum_{k=1}^{K} \mathcal{X}_t^{(k)}, \tag{6.63}$$

where $\boldsymbol{A}^{(k)}$ is the adjacency matrix of the k-th sub-graph. The decomposition prediction problem is then defined as:

$$\hat{\mathbf{Y}}_{>t}^{(k)} = f_d\left(\mathcal{X}_t^{(k)}, G^{(k)}, \Theta^{(k)}\right) \quad s.t. \; \hat{\mathbf{Y}}_{>t}^{(g)} = \sum_{k=1}^{K} \hat{\mathbf{Y}}_{>t}^{(k)}, \tag{6.64}$$

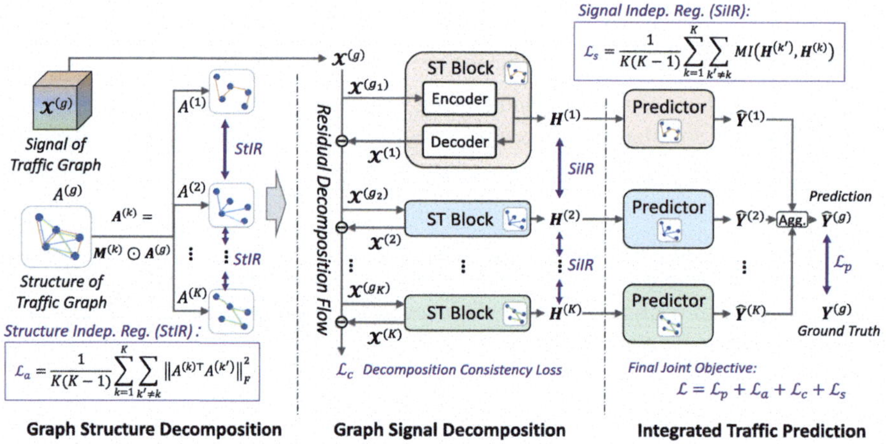

Fig. 6.2 Overview architecture of the TGDL framework

where $\Theta^{(k)}$ represents the parameters of the k-th prediction function.

In this chapter, the TGDL framework implements decomposition prediction in three steps (see Fig. 6.2 for details): the graph structure decomposition module, the graph signal decomposition module, and the traffic prediction integration module, which are introduced sequentially below.

6.3.2 *Traffic Graph Structure Decomposition*

The TGDL framework first applies Structure-Independent Decomposition to partition the original input graph $G^{(g)}$ into multiple sub-graphs. For the k-th sub-graph, its node set is identical to the original graph, while its corresponding adjacency matrix $\mathbf{A}^{(k)}$ is calculated as:

$$\mathbf{A}^{(k)} = \mathbf{M}^{(k)} \odot \mathbf{A}^{(g)}, \tag{6.65}$$

where $\odot$ denotes the Hadamard Product (element-wise multiplication). $\mathbf{M}^{(k)}$ is a mask matrix where the element at the i-th row and j-th column, $m_{ij}^{(k)} \in [0, 1]$, represents the interaction relationship between nodes in the k-th sub-graph.

To ensure that the adjacency matrices of all sub-graphs can reconstruct the original graph, satisfying Eq. (6.63) in Definition 6.7, this chapter employs the SoftMax function to compute $m_{ij}^{(k)}$ as follows:

$$m_{ij}^{(k)} = \frac{\exp\left(w_{ij}^{(k)}\right)}{\sum_{k'=1}^{K} \exp\left(w_{ij}^{(k')}\right)}, \tag{6.66}$$

where $w_{ij}^{(k)}$ are learnable dynamic parameters. This formula ensures that for each element in the mask matrices of different sub-graphs, the relationship $\sum_{k=1}^{K} m_{ij}^{(k)} = 1$ holds. Consequently, for every edge in the original graph structure, $\sum_{k=1}^{K} a_{ij}^{(k)} = a_{ij}^{(g)}$ applies.

According to the Component Independence (CI) Principle, an effective decomposition algorithm should ensure that decomposed sub-components are independent of each other. Therefore, structure independence regularization is introduced here to constrain the sub-graph adjacency matrices:

$$\mathcal{L}_a = \frac{1}{K(K-1)} \sum_{k=1}^{K} \sum_{k' \neq k} \left\| \boldsymbol{A}^{(k)\top} \boldsymbol{A}^{(k')} \right\|_F^2, \tag{6.67}$$

where $\|\cdot\|_F$ denotes the Frobenius norm. By minimizing $\mathcal{L}_a$, the orthogonality between any pair of sub-graph adjacency matrices is enhanced, thereby ensuring mutual independence among sub-graph structures.

6.3.3 *Traffic Graph Signal Decomposition*

This section introduces the graph signal decomposition module in the TGDL model, which primarily employs a Residual Decomposition Flow (RDF) to iteratively extract the signal component corresponding to each sub-graph structure from the original graph, subsequently utilizing independence regularization to ensure independence among sub-graph signals.

6.3.3.1 Residual Decomposition Flow

Let $\mathcal{X}_t^{(g)}$ denote the signal of the original graph at time step t. For the first sub-graph $G^{(1)}$, its signal at time t is extracted from the input graph signal $\mathcal{X}_t^{(g)}$ via a Spatiotemporal Block (ST Block) structure, as follows:

$$\mathcal{X}_t^{(1)} = \text{STBlock}\left(\mathcal{X}_t^{(g)}, \boldsymbol{A}^{(1)}, \Theta_b^{(1)}\right), \tag{6.68}$$

where $\mathcal{X}_t^{(1)} = \{\mathbf{X}_{t-T}^{(1)}, \ldots, \mathbf{X}_t^{(1)}\}$ is the signal of $G^{(1)}$, and $\Theta_b^{(1)}$ represents learnable model parameters. The function $\text{STBlock}(\cdot)$ is implemented using Graph Neural Networks, with specific details provided in the next subsection.

Then, by subtracting the signal $\mathcal{X}_t^{(1)}$ corresponding to sub-graph $G^{(1)}$ from the original graph signal $\mathcal{X}_t^{(g)}$, the residual signal is obtained:

$$\mathcal{X}_t^{(g_2)} = \mathcal{X}_t^{(g)} - \mathcal{X}_t^{(1)}. \tag{6.69}$$

Subsequently, an ST Block is used again to extract the signal of the second sub-graph $G^{(2)}$ from $\mathcal{X}_t^{(g_2)}$, following a process similar to Eq. (6.68):

$$\mathcal{X}_t^{(2)} = \text{STBlock}\left(\mathcal{X}_t^{(g_2)}, \boldsymbol{A}^{(2)}, \Theta_b^{(2)}\right). \tag{6.70}$$

Given K sub-graph structures, by iteratively applying Eqs. (6.68) and (6.69), a Residual Decomposition Flow can be constructed. The input signal for the k-th sub-graph is:

$$\mathcal{X}_t^{(g_k)} = \mathcal{X}_t^{(g_{k-1})} - \mathcal{X}_t^{(k-1)}. \tag{6.71}$$

The signal component of the k-th sub-graph extracted via the Residual Decomposition Flow is:

$$\mathcal{X}_t^{(k)} = \text{STBlock}\left(\mathcal{X}_t^{(g_k)}, \boldsymbol{A}^{(k)}, \Theta_b^{(k)}\right). \tag{6.72}$$

To maintain notation consistency, let $\mathcal{X}_t^{(g_1)} = \mathcal{X}_t^{(g)}$. Through the iterative algorithm process defined by Eqs. (6.71) and (6.72), the original signal $\mathcal{X}_t^{(g)}$ is decomposed into a series of sub-graph signals, namely $\{\mathcal{X}_t^{(1)}, \ldots, \mathcal{X}_t^{(K)}\}$.

According to the definition of decomposition prediction in Eq. (6.63) within Definition 6.7, the sum of sub-graph signals should equal the original signal. Therefore, a Decomposition Consistency Loss is introduced to constrain the sub-graph signals:

$$\mathcal{L}_c = \left\| \mathcal{X}_t^{(g)} - \sum_{k=1}^{K} \mathcal{X}_t^{(k)} \right\|_F^2. \tag{6.73}$$

In the Residual Decomposition Flow described by Eq. (6.71), the signal component of each sub-graph is sequentially removed from the original signal. Thus, sub-graph signals extracted by Eq. (6.72) are decoupled from each other, satisfying the CI Principle to a certain extent.

6.3.3.2 Spatiotemporal Block

In the TGDL framework, Graph Neural Networks with an "Encoder-Decoder" architecture can be utilized to implement the ST Block, i.e., the STBlock$(\cdot)$ function in Eq. (6.72). Specifically, given the residual graph signal $\mathcal{X}_t^{(g_k)}$ of the k-th sub-graph, the ST Block uses an encoder to extract signal representations:

$$\mathbf{H}_t^{(k)} = \text{Encoder}\left(\mathcal{X}_t^{(g_k)}, \boldsymbol{A}^{(k)}, \Theta_e^{(k)}\right), \tag{6.74}$$

where $\mathbf{H}_t^{(k)} \in \mathbb{R}^{|\mathcal{V}| \times D}$ is the representation matrix generated by the signal of sub-graph k. Each row of this matrix $\mathbf{h}_i^{(k)} \in \mathbb{R}^D$ corresponds to the representation of node v_i, where D is the dimension of the representation vector. Additionally, $\Theta_e^{(k)}$ denotes the learnable parameters of the encoder module.

Next, for the k-th sub-graph signal, the ST Block employs a decoder to transform the representation $\mathbf{H}^{(k)}$ into:

$$\mathcal{X}_t^{(k)} = \text{Decoder}\left(\mathbf{H}_t^{(k)}, A^{(k)}, \Theta_d^{(k)}\right), \tag{6.75}$$

where $\mathcal{X}_t^{(k)}$ is the decomposed signal corresponding to sub-graph $G^{(k)}$, and $\Theta_d^{(k)}$ represents the learnable parameters of the decoder. In this manner, the encoder in Eq. (6.74) and the decoder in Eq. (6.75) constitute the STBlock$(\cdot)$ function in Eq. (6.72). The encoder and decoder functions can be implemented using various traffic graph data-based encoder-decoder models from existing literature [12, 30, 33].

6.3.3.3 Signal Independence Regularization

According to the CI Principle in Theorem 6.4, signal components of different sub-graphs should be mutually independent. In the Residual Decomposition Algorithm process formed by Eqs. (6.71) and (6.72), the original signal is processed into a series of decoupled sub-graph signals. To further strengthen the independence among sub-graph signals, Signal Independence Regularization (SIR) is introduced to constrain the relationships between signal representation matrices generated by ST Blocks.

Given signal representation matrices $\mathbf{H}^{(k)}$ $(k \in \{1, \ldots K\})$ generated by Eq. (6.74), SIR can be defined as:

$$\mathcal{L}_s = \frac{1}{K(K-1)} \sum_{k=1}^{K} \sum_{k' \neq k} \text{MI}\left(\mathbf{H}^{(k')}, \mathbf{H}^{(k)}\right), \tag{6.76}$$

where $\text{MI}(\mathbf{H}^{(k')}, \mathbf{H}^{(k)})$ denotes the Mutual Information (MI) between signal representations of sub-graphs k' and k, defined as:

$$\text{MI}\left(\mathbf{H}^{(k')}, \mathbf{H}^{(k)}\right) = \mathbb{E}\left[\log \frac{\Pr\left(\mathbf{H}^{(k')}, \mathbf{H}^{(k)}\right)}{\Pr\left(\mathbf{H}^{(k')}\right) \cdot \Pr\left(\mathbf{H}^{(k)}\right)}\right], \tag{6.77}$$

In this formula, $\Pr(\mathbf{H}^{(k)})$ and $\Pr(\mathbf{H}^{(k')})$ represent the marginal distributions of signals $\mathbf{H}^{(k)}$ and $\mathbf{H}^{(k')}$ respectively, while $\Pr(\mathbf{H}^{(k')}, \mathbf{H}^{(k)})$ represents their joint distribution.

Since closed-form expressions for the joint and marginal distributions of $\mathbf{H}^{(k')}$ and $\mathbf{H}^{(k)}$ are difficult to solve, direct calculation of MI in Eq. (6.77) is unfeasible. Given the objective to minimize the MI between $\mathbf{H}^{(k')}$ and $\mathbf{H}^{(k)}$, this chapter adopts the CLUB method [3] to approximate the Upper Bound (UB) of MI, calculated as:

$$\widehat{\mathrm{MI}}_{\mathrm{UB}}\left(\mathbf{H}^{(k')},\mathbf{H}^{(k)}\right)=\frac{1}{N}\sum_{n=1}^{N}\left[\log q_\theta\left(\mathbf{H}_n^{(k')}|\mathbf{H}_n^{(k)}\right)-\frac{1}{N}\sum_{n'=1}^{N}\log q_\theta\left(\mathbf{H}_{n'}^{(k')}|\mathbf{H}_n^{(k)}\right)\right], \tag{6.78}$$

where N is the number of samples, and $\mathbf{H}_i^{(k)}$ and $\mathbf{H}_i^{(k')}$ denote the signal representation matrices for sub-graphs k and k' corresponding to the i-th sample, respectively. In Eq. (6.78), $q_\theta(\mathbf{H}^{(k')}|\mathbf{H}^{(k)})$ represents the variational approximation of the conditional probability $\Pr(\mathbf{H}^{(k')}|\mathbf{H}^{(k)})$. Without loss of generality, it is assumed that this variational approximation follows a Gaussian distribution $\mathcal{N}(\mu_{\mathbf{H}^{(k)}},\sigma^2_{\mathbf{H}^{(k)}})$, i.e.,

$$q_\theta\left(\mathbf{H}^{(k')}|\mathbf{H}^{(k)}\right)=-\frac{1}{\sqrt{2\pi}\sigma_{\mathbf{H}^{(k)}}}\exp\left(-\frac{(x-\mu_{\mathbf{H}^{(k)}})^2}{2\sigma^2_{\mathbf{H}^{(k)}}}\right). \tag{6.79}$$

For the mean $\mu_{\mathbf{H}^{(k)}}$ and variance $\sigma^2_{\mathbf{H}^{(k)}}$ in the Gaussian distribution, this chapter employs a Multi-Layer Perceptron (MLP) for estimation, taking $\mathbf{H}^{(k)}$ as input, such that:

$$\left\{\mu_{\mathbf{H}^{(k)}},\sigma^2_{\mathbf{H}^{(k)}}\right\}=\mathrm{MLP}\left(\mathbf{H}^{(k)},\Theta_{\mathrm{mlp}}\right), \tag{6.80}$$

where Θ_{mlp} represents all learnable parameters in the MLP.

Finally, substituting the estimated MI $\widehat{\mathrm{MI}}_{\mathrm{UB}}(\mathbf{H}^{(k')},\mathbf{H}^{(k)})$ from Eq. (6.78) into Eq. (6.76) implements signal independence regularization. By minimizing the mutual information between $\mathbf{H}^{(k')}$ and $\mathbf{H}^{(k)}$, independence between signals of sub-graphs k and k' is enhanced, thereby ensuring the Residual Decomposition Flow complies with the CI Principle derived in Theorem 6.4.

6.3.4 Traffic Prediction Integration

The goal of the traffic prediction integration module is to forecast future traffic states based on the signal representations of each sub-graph and then aggregate the prediction results from all sub-graphs to obtain the final prediction as the solution to the original problem.

For each sub-graph $G^{(k)}$, this chapter inputs its signal representation $\mathbf{H}^{(k)}$ into a predictor to forecast future traffic states:

$$\hat{\mathbf{Y}}_{>t}^{(k)}=\mathrm{Predictor}\left(\mathbf{H}_t^{(k)},\mathbf{A}^{(k)},\Theta_p^{(k)}\right). \tag{6.81}$$

Here, $\mathrm{Predictor}(\cdot)$ is a prediction function with a model structure similar to the decoder in Eq. (6.75), but with different output $\hat{\mathbf{Y}}_{>t}^{(k)}$ and learnable parameters $\Theta_p^{(k)}$. Subsequently, the final forecast for future traffic states is obtained by summing the prediction results from each sub-graph:

$$\hat{\mathbf{Y}}_{>t}^{(g)} = \sum_{k=1}^{K} \hat{\mathbf{Y}}_{>t}^{(k)}. \tag{6.82}$$

To measure the error of the final prediction result, Mean Square Error (MSE), commonly used in prediction tasks, is adopted as the loss function, calculated as:

$$\mathcal{L}_p = \left\| \mathbf{Y}_{>t} - \hat{\mathbf{Y}}_{>t}^{(g)} \right\|_F^2. \tag{6.83}$$

In the proposed TGDL framework, the decoder in Eq. (6.75) and the predictor in Eq. (6.81) share the same model structure but possess different model parameters. Crucially, this "Encoder-Decoder (Predictor)" structure can be implemented using any "Encoder-Decoder" based traffic graph prediction model [12, 30, 33], which are common in spatiotemporal forecasting. For distinction, spatiotemporal prediction models implemented based on "Encoder-Decoder" structures are referred to as *base models* in subsequent content. TGDL can be integrated into these base models to enhance their performance across various contexts. Thus, TGDL can be regarded as a *model-agnostic decomposition forecasting framework*.

6.3.5 *Model Training Process*

The objective function for model training consists of three parts: (i) the loss function $\mathcal{L}_p$ in Eq. (6.83), used to minimize traffic prediction error; (ii) the decomposition consistency loss $\mathcal{L}_c$ in Eq. (6.73), used to constrain sub-graph signals to meet decomposition prediction requirements; and (iii) independence regularization terms $\mathcal{L}_a$ and $\mathcal{L}_s$ in Eqs. (6.67) and (6.76), used to ensure the traffic graph decomposition process adheres to the CI Principle derived in Theorem 6.4. The final joint objective function is defined as:

$$\mathcal{L} = \gamma_p \mathcal{L}_p + \gamma_a \mathcal{L}_a + \gamma_c \mathcal{L}_c + \gamma_s \mathcal{L}_s. \tag{6.84}$$

To avoid manually selecting hyperparameters, the weights $\gamma_p, \gamma_a, \gamma_c, \gamma_s$ are set through a Dynamic Weight Averaging (DWA) strategy [17], with an initial value of 1. DWA dynamically adjusts the weight of each loss term based on its change rate during training to balance the loss changes of all tasks. This prevents any single task from dominating, promoting balanced learning across all tasks.

6.4 Empirical Results and Insights

This section presents the empirical results and key insights of the proposed methods. It begins with the evaluation setup, followed by an overall comparison and analysis of results. The section then reports ablation studies on key modules and concludes with an in-depth analysis of model behavior.

6.4.1 Evaluation Setup

This subsection describes the experimental settings used for empirical evaluation, including the dataset characteristics, the evaluation metrics, baselines for comparison, and key implementation details.

6.4.1.1 Data Description

To evaluate the effectiveness of the TGDL framework, experiments were conducted on four real-world traffic datasets:

- NYCBike: A public traffic flow dataset collected from New York City, containing bicycle rental data from April 1, 2014, to September 30, 2014 [34]. The dataset partitions the city into a 16×8 grid, mapping bicycle rental records to traffic flow between regions. Data is recorded hourly, totaling 4,392 samples.
- NYCTaxi: Another public dataset from New York City, containing taxi GPS data from January 1, 2015, to March 1, 2015. The city is partitioned into a 20×10 grid, mapping taxi trajectories to traffic flow between regions [32]. Traffic flow is recorded every 30 min, totaling 2,880 samples.
- PEMSD4: A traffic dataset collected by Caltrans [2], containing data from 307 freeway sensors in the San Francisco Bay Area over two months (January 1, 2018, to February 28, 2018) [10]. Original data is recorded every 5 min. To reduce noise impact, the dataset used in experiments aggregates traffic data at 30-minute intervals, yielding 2,832 samples.
- PEMSD8: Another dataset from Caltrans, containing data from 170 sensors in District 8 over two months (July 1, 2016, to August 31, 2016) [10]. Similarly, this dataset aggregates original 5-minute interval data into 30-minute intervals, yielding 2,976 samples.

For each prediction sample at time step t, two types of historical data are used as input: (i) data from the 4 h prior to time t; and (ii) data from 2 h around the time slots $t - T_{\text{day}}$, $t - 2 \times T_{\text{day}}$, and $t - 3 \times T_{\text{day}}$, where T_{day} is the number of time intervals in a day. It is noted that the second type of input is intended to incorporate periodic information. A gap-free sliding window strategy is used to generate samples, and each dataset is divided into training, validation, and test sets in a 7:1:2 ratio.
Traffic Graph Structure

The graph structure is represented by an adjacency matrix $\boldsymbol{A}$, where a_{ij} is the element in the i-th row and j-th column. Following standard practices in existing literature [15], a thresholded Gaussian kernel is used to construct $\boldsymbol{A}$ as follows:

$$a_{ij} = \begin{cases} \exp\left(-\frac{\mathcal{D}(v_i,v_j)^2}{\sigma^2}\right), & \mathcal{D}(v_i, v_j) \leq r, \\ 0, & \mathcal{D}(v_i, v_j) > r. \end{cases} \tag{6.85}$$

Here, $\mathcal{D}(v_i, v_j)$ represents the distance between node v_i and v_j, σ is the standard deviation of distances, and the threshold r is set to 0.6. For PEMSD4 and PEMSD8, $\mathcal{D}(v_i, v_j)$ is calculated based on the actual road network distance between sensors. For NYCBike and NYCTaxi, due to the lack of road network information, the Dynamic Time Warping (DTW) algorithm is employed, using traffic time series between nodes as input for distance measurement, i.e., $\mathcal{D}(v_i, v_j) = \text{DTW}(\mathbf{x}_i, \mathbf{x}_j)$.

Task Description

For NYCBike and NYCTaxi, the datasets provide inflow and outflow counts for each region. This enables direct modeling of spatially distributed inflow/outflow patterns. Following prior works [12, 32, 34], we adopt a single-step prediction setting, where the model predicts both inflow and outflow at the next time step. For PEMSD4 and PEMSD8, the datasets contain aggregated traffic volume at each sensor location, without separate inflow and outflow information. Therefore, consistent with common practice in existing works [10, 11, 29], we adopt a multi-step prediction setting, where the model forecasts the total traffic flow from 30 min to 2 h ahead (i.e. 1–4 future steps). To validate that our framework generalizes across different prediction settings, we have also included multi-step prediction results on the NYCBike and NYCTaxi datasets in the Supplementary Materials.

In the theoretical derivation, we assume the traffic state Y is a discrete variable. Empirically, our experiment focuses on traffic flow data, whose smallest unit is 1. This allows them to be treated as discrete variables without further discretization, thus satisfying the "discrete value" assumption. For inherently continuous traffic states like speed, discretization is done based on practical needs, such as categorizing speed into levels for congestion scenarios [25].

6.4.1.2 Experimental Scenarios and Baselines

To evaluate the efficacy of the proposed TGDL framework, two experimental scenarios were designed, including experiments with different base models and comparison with decomposition baselines. They are introduced in the following.

Experiments with Different Base Models

In the first scenario, various GNN-based "Encoder-Decoder" spatiotemporal prediction models serve as base models for TGDL, comparing the performance of TGDL-enhanced models against original base models.

This scenario verifies the effectiveness of TGDL for base models and demonstrates its robustness and stability across different models. Baseline methods for this scenario include five state-of-the-art GNN-based spatiotemporal prediction models.

- STGCN [33]: A graph convolution-based model using 1D convolution to capture temporal correlations.

- GWNet [30]: Combines diffusion graph convolution with dilated causal convolution to capture spatiotemporal dependencies.
- MTGNN [29]: Integrates multi-order graph convolution and temporal convolution for multivariate time series forecasting.
- MSDR [16]: A recurrent neural network variant that combines graph convolution in each temporal modeling unit for spatiotemporal prediction, explicitly utilizing hidden states from multiple historical steps as input.
- STSSL [12]: A self-supervised spatiotemporal prediction model that designs two self-supervised auxiliary tasks on top of classic models to model spatiotemporal heterogeneity, thereby obtaining more effective spatiotemporal representations.

Comparison with Decomposition Baselines
In the second experimental scenario, TGDL is compared with decomposition-based prediction baselines across three categories. The first category is mainly two-stage decomposition prediction approaches.

- PTD-XFMR [35]: This is a classic two-stage method, where the decomposition is performed using Periodic-Trend Decomposition (PTD), and forecasting is conducted using sequence learning models such as LSTM. To enhance this model, we replace the LSTM with the powerful Transformer (XFMR) model.
- EMD-XFMR [18]: The decomposition step is performed by the two-level Empirical Mode Decomposition (EMD). The prediction step is realized by GRU, which is replaced with Transformer (XFMR) for stronger modeling capacity.

The second category includes state-of-the-art models with signal decomposition.

- CoST [28]: It uses contrastive learning to decompose seasonal and trend hidden representations for time series forecasting.
- D2STGNN [21]: It decomposes data into diffusion and non-diffusion signals for traffic prediction.
- STWave [7]: It decomposes complex traffic data into stable trends and volatile events to improve prediction performance.
- SCNN [5]: It iteratively decomposes input data into long-term, seasonal, short-term, and co-evolutionary components for more accurate spatiotemporal prediction.

The third category comprises prediction models with both structure and signal decomposition. As existing decomposition prediction models primarily focus on signal decomposition, this study applies TGDL's signal decomposition alongside two state-of-the-art graph structure decomposition methods as baselines for this category.

- TGDL-METIS: This variant replaces the graph structure decomposition module in TGDL with METIS [13], a hierarchical node-based graph partitioning algorithm.
- TGDL-DBH: The model replaces the graph structure decomposition module with Degree-Based Hashing (DBH) [31], an edge-based graph decomposition method.

Moreover, we include two types of classic non-deep models as a basic performance reference.

- HA: It models spatiotemporal data as a seasonal process, predicting by averaging historical data from target times.
- SVR: It uses support vector regression with a linear kernel, representing a classic non-deep learning approach.

6.4.1.3 Implementation Details

The TGDL model was implemented using Python 3.9 and PyTorch 1.12.0. Model training utilized the Adam optimizer with an initial learning rate of 0.001. The number of sub-graphs was selected within the range [2, 15] for all datasets, with optimal numbers for NYCBike, NYCTaxi, PEMSD4, and PEMSD8 being 6, 6, 7, and 12, respectively. Hidden layer dimensions were set according to the original implementations in the open-source code of each base model.

6.4.2 Overall Results and Comparative Analysis

Consistent with the experimental scenario setup, we evaluated the experimental results under both configurations.

6.4.2.1 Experiments with Different Base Models

Results for this scenario are summarized in Tables 6.2 and 6.3. Best results are bolded, and second-best results are underlined. "In" and "Out" denote Inflow and Outflow, while "30 min" and "2 h" correspond to prediction lengths of 30 min (1 step) and 2 h (4 steps). TGDL-enhanced base models are denoted as "Base Model+", e.g. STGCN+ for TGDL-enhanced STGCN.

To assess prediction accuracy, three common metrics are used: Mean Absolute Error (MAE), Root Mean Square Error (RMSE), and Mean Absolute Percentage Error (MAPE), with MAPE in %. All experiments were run five times with different random seeds, and the average performance is reported. Additionally, Tables 6.2 and 6.3 present the Relative Error Reduction Rate (RERR) of TGDL compared to original base models.

Based on experimental results, we make the following observations.

Firstly, for all base models, the performance of TGDL-enhanced models is significantly better than the corresponding original base models in every case, with an average error reduction rate of 19.37%, 17.62%, and 19.49% in MAE, RMSE, and MAPE, respectively. This validates the effectiveness of TGDL as a framework to enhance the predictive capabilities of various GNN-based models. Such effectiveness can also be generalized to more traffic data types and even other spatiotemporal domains beyond traffic.

Table 6.2 Comparison of enhancement effects of TGDL framework on different base models over NYC datasets

Dataset	NYCBike						NYCTaxi					
Metric	MAE ↓		RMSE ↓		MAPE ↓		MAE ↓		RMSE ↓		MAPE ↓	
Type	In	Out	In	Out	In	Out	In	Out	In	Out	In	Out
STGCN	5.26	5.53	7.65	8.23	22.58	23.13	12.87	10.93	23.09	23.62	17.25	17.23
STGCN+	<u>5.07</u>	5.36	<u>7.36</u>	<u>7.95</u>	<u>21.77</u>	<u>22.55</u>	11.84	<u>9.68</u>	21.17	17.61	17.13	<u>16.80</u>
RERR	−3.61%	−3.07%	−3.79%	−3.40%	−3.59%	−2.51%	−8.00%	−11.44%	−8.32%	−25.45%	−0.70%	−2.50%
GWNet	5.15	5.43	7.38	8.01	22.67	23.24	12.70	10.39	22.48	18.29	17.45	17.79
GWNet+	**4.97**	**5.25**	**7.12**	**7.75**	**21.61**	**22.28**	**11.53**	**9.32**	<u>20.97</u>	**16.15**	**15.78**	**16.28**
RERR	−3.50%	−3.31%	−3.52%	−3.25%	−4.68%	−4.13%	−9.21%	−10.30%	−6.72%	−11.70%	−9.57%	−8.49%
MTGNN	5.26	5.57	7.68	8.31	22.77	23.83	12.71	10.19	22.37	18.00	17.03	17.71
MTGNN+	5.12	5.42	7.46	8.12	22.28	23.18	11.94	9.71	21.07	<u>16.90</u>	16.38	16.98
RERR	−2.66%	−2.69%	−2.86%	−2.29%	−2.15%	−2.73%	−6.06%	−4.71%	−5.81%	−6.11%	−3.82%	−4.12%
MSDR	5.65	6.00	8.26	8.99	24.54	25.89	18.49	15.89	33.7	33.17	25.27	25.78
MSDR+	5.25	5.55	7.68	8.31	23.22	23.86	14.72	11.53	26.31	21.44	20.61	20.37
RERR	−7.08%	−7.50%	−7.02%	−7.56%	−5.38%	−7.84%	−20.39%	−27.44%	−21.93%	−35.36%	−18.44%	−20.99%
STSSL	5.18	5.44	7.56	8.15	22.53	23.22	12.19	10.00	21.70	19.05	16.51	17.21
STSSL+	5.10	<u>5.35</u>	7.45	8.03	22.36	22.99	<u>11.83</u>	9.74	**20.86**	17.06	<u>16.36</u>	17.06
RERR	−1.54%	−1.65%	−1.46%	−1.47%	−0.76%	−0.99%	−2.95%	−2.60%	−3.87%	−10.45%	−0.91%	−0.87%

Table 6.3 Comparison of enhancement effects of TGDL framework on different base models on PEMSD datasets

Dataset	PEMSD4						PEMSD8					
Metric	MAE ↓		RMSE ↓		MAPE ↓		MAE ↓		RMSE ↓		MAPE ↓	
Type	30 min	2 h	30 min	2 h	30 min	2 h	30 min	2 h	30 min	2 h	30 min	2 h
STGCN	28.88	30.07	42.36	44.20	22.21	20.96	23.61	26.96	34.69	40.55	14.65	15.35
STGCN+	12.45	15.87	20.00	26.37	8.11	9.92	9.75	12.45	15.97	21.46	5.68	6.98
RERR	–56.89%	–47.22%	–52.79%	–40.34%	–63.48%	–52.67%	–58.70%	–53.82%	–53.96%	–47.08%	–61.23%	–54.53%
GWNet	12.52	18.39	19.96	29.18	8.05	11.80	9.83	14.59	16.20	23.46	5.50	8.48
GWNet+	**11.22**	**14.58**	**18.27**	**24.61**	7.34	9.37	8.62	**11.36**	**14.68**	**20.30**	**4.83**	**6.52**
RERR	–10.38%	–20.72%	–8.47%	–15.66%	–8.82%	–20.59%	–12.31%	–22.14%	–9.38%	–13.47%	–12.18%	–23.11%
MTGNN	12.28	18.38	19.84	30.40	7.70	11.23	9.65	13.55	15.95	23.11	5.50	7.78
MTGNN+	11.29	15.39	18.62	25.84	**6.88**	**9.30**	**8.60**	11.43	14.82	20.31	4.95	6.89
RERR	–8.06%	–16.27%	–6.15%	–15.00%	–10.65%	–17.19%	–10.88%	–15.65%	–7.08%	–12.12%	–10.00%	–11.44%
MSDR	27.75	25.66	44.44	38.81	19.01	18.25	23.21	23.56	39.46	34.75	14.03	14.69
MSDR+	19.09	20.59	31.27	33.17	12.96	14.15	17.65	17.98	34.72	30.62	9.91	10.42
RERR	–31.21%	–19.76%	–29.64%	–14.53%	–31.83%	–22.47%	–23.96%	–23.68%	–12.01%	–11.88%	–29.37%	–29.07%
STSSL	28.87	29.22	41.85	43.62	21.48	19.90	22.95	27.18	34.26	40.01	14.74	16.03
STSSL+	12.90	16.17	20.51	26.81	8.22	10.11	10.50	13.77	16.89	22.91	6.30	8.40
RERR	–55.32%	–44.66%	–50.99%	–38.54%	–61.73%	–49.20%	–54.25%	–49.34%	–50.70%	–42.74%	–57.26%	–47.60%

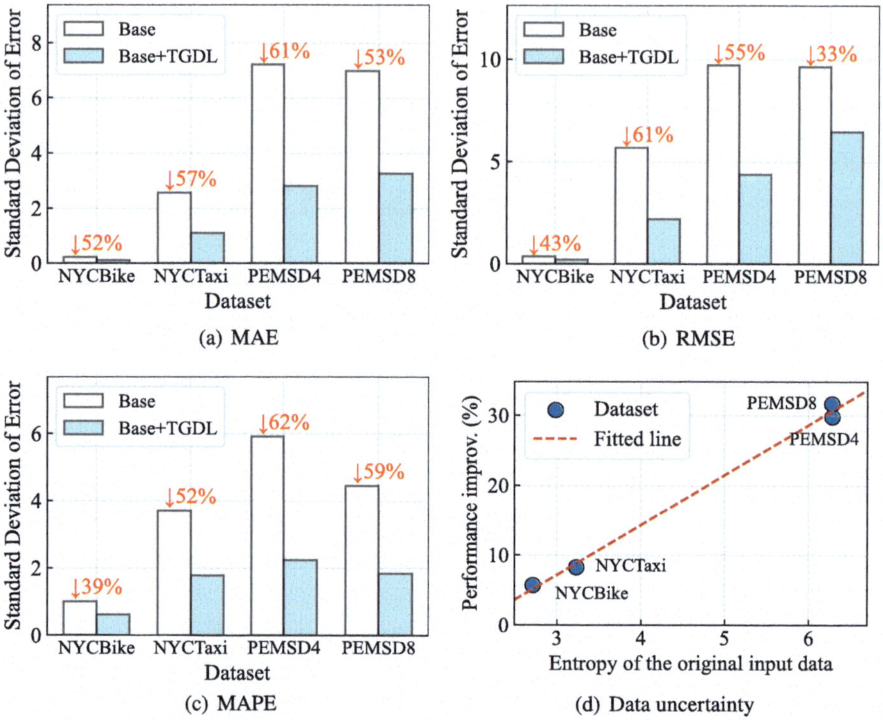

Fig. 6.3 Model performance variance before and after TGDL enhancement and impact of data uncertainty

Secondly, GWNet+ outperforms other enhanced models for two main reasons: (1) Its base version, GWNet, is already a powerful model capable of capturing complex spatiotemporal dependencies, thereby providing a strong foundation for accurate prediction. This results in a lower model-induced (MI) error. (2) With the integration of TGDL, which reduces data-induced (DI) error, GWNet's modeling capabilities are fully leveraged. Consequently, GWNet+ achieves the lowest total error.

Thirdly, although the enhanced GWNet achieves the best overall performance, the relative improvement brought by TGDL is more pronounced in simpler models such as STGCN. This finding demonstrates that TGDL effectively reduces the intrinsic difficulty of the prediction task, thereby diminishing the necessity for highly complex models. As a result, TGDL enables lightweight models to achieve competitive accuracy, making them more suitable for practical applications where computational resources may be limited.

Fourthly, for base models with relatively lower original performance, integrating the TGDL framework results in a greater enhancement, reducing performance differences between base models. Figure 6.3a–c provide additional insight into this phenomenon, showing that the reduced performance standard deviations for TGDL-enhanced models. This indicates that with decomposition, the complexity of the subproblem decreases, narrowing the performance gap across different models.

Finally, the performance gains for the PEMS datasets are significantly greater than those for the NYC datasets because datasets with higher uncertainty, measured by the entropy $H(Y|X)$, tend to benefit more from our TGDL. Figure 6.3d illustrates the relationship between TGDL's average performance gains and data complexity. As shown, datasets like PEMSD4 and PEMSD8, which have higher entropy, achieve larger improvements. This disparity arises from dataset composition: PEMS datasets include multiple vehicle types resulting in more complex patterns. Therefore, calculating the entropy of a dataset before applying TGDL can help identify where the model is likely to provide the most significant benefits.

6.4.2.2 Comparison with Decomposition Approaches

Table 6.4 presents the results of TGDL in comparison to other decomposition-enhanced baselines. Here, the base model for TGDL is STGCN, chosen as the most widely cited base model. The following observations can be made.

Firstly, the classic non-deep methods, HA and SVR, yield suboptimal results due to their limited expressive power. In contrast, deep learning models, particularly those enhanced with TGDL, can capture complex spatiotemporal relationships, resulting in more accurate predictions.

Secondly, the two-stage decomposition prediction approaches generally underperform compared to end-to-end paradigms, primarily due to the objective inconsistency between the decomposition and prediction stages.

Thirdly, among the baselines with signal decomposition, SCNN achieves the best average performance. Compared to other signal decomposition baselines, SCNN decomposes the input data into a wider variety of component types, increasing the likelihood of their independence from each other. This approach aligns with the CI Principle, benefiting performance. However, none of these baselines include explicit mechanisms to ensure maximal independence of decomposition components, leading to lower performance than TGDL and its variants.

Fourthly, the variants TGDL-METIS and TGDL-DBH outperform other representative baselines, highlighting that graph structure decomposition, in addition to signal decomposition, is also crucial for ensuring component independence and reducing prediction error. Furthermore, the TGDL framework in these variants promotes component independence, which contributes to improved performance.

Finally, our proposed TGDL consistently outperforms all other baselines across all four datasets, verifying the effectiveness of our method. The performance of TGDL-METIS and TGDL-DBH is slightly lower than that of TGDL, likely because these variants separately optimize the structure and signal decomposition modules, whereas TGDL jointly optimizes both modules in an end-to-end manner.

Table 6.4 Comparison of results for different decomposition prediction models

Dataset	NYCBike						NYCTaxi					
Metric	MAE ↓		RMSE ↓		MAPE ↓		MAE ↓		RMSE ↓		MAPE ↓	
Type	In	Out	In	Out	In	Out	In	Out	In	Out	In	Out
HA	10.85	11.09	16.91	17.50	45.68	46.98	36.76	28.51	66.87	56.68	46.72	44.02
SVR	8.48	8.59	13.67	14.31	34.63	34.63	35.81	28.65	68.91	58.05	45.62	43.47
PTD-XFMR	7.77	8.12	11.21	12.44	33.43	35.12	17.53	13.56	30.89	23.86	22.79	22.06
EMD-XFMR	7.18	7.83	10.61	11.57	31.35	33.73	17.15	12.62	29.19	22.32	21.32	21.38
CoST	6.58	7.06	9.47	10.29	30.10	31.55	17.45	13.73	30.27	24.96	24.70	23.66
D2STGNN	5.63	5.95	8.07	8.80	24.74	25.73	16.50	13.60	29.24	25.29	20.07	19.20
STWave	7.11	7.63	10.30	11.33	30.39	32.60	16.38	12.55	28.89	22.64	21.31	20.78
SCNN	6.10	6.59	8.79	9.87	27.77	29.31	14.32	11.39	24.67	19.63	21.03	21.23
TGDL-METIS	5.18	5.46	7.62	8.09	22.11	23.09	11.94	9.83	21.62	18.72	18.81	17.25
TGDL-DBH	5.16	5.50	7.50	8.15	22.10	23.09	12.35	10.01	21.81	18.30	18.20	18.54
TGDL	**5.07**	**5.36**	**7.36**	**7.95**	**21.77**	**22.55**	**11.84**	**9.68**	**21.17**	**17.61**	**17.13**	**16.80**
Dataset	PEMSD4						PEMSD8					
Metric	MAE ↓		RMSE ↓		MAPE ↓		MAE ↓		RMSE ↓		MAPE ↓	
Type	30 min	2 h	30 min	2 h	30 min	2 h	30 min	2 h	30 min	2 h	30 min	2 h
HA	40.70	63.33	61.78	89.72	32.80	53.44	37.03	55.74	54.15	77.98	24.84	36.67
SVR	40.80	47.59	61.57	71.09	32.75	34.17	36.83	41.43	54.88	59.99	26.20	26.67
PTD-XFMR	20.19	28.18	30.19	42.24	12.25	18.37	18.22	28.52	26.92	41.67	10.17	16.42
EMD-XFMR	19.27	26.95	28.55	39.81	11.76	17.07	17.75	27.02	25.18	41.37	9.86	15.58
CoST	21.00	24.59	32.71	37.07	14.79	17.96	20.87	25.04	32.78	37.90	12.00	14.78
D2STGNN	17.07	22.91	24.64	34.78	15.21	16.65	15.51	23.93	23.25	34.98	12.37	21.85
STWave	18.46	26.03	28.03	39.79	11.51	16.89	16.92	26.20	25.09	39.45	9.62	15.46
SCNN	13.79	21.72	22.07	34.83	8.37	13.74	10.78	16.82	17.19	26.20	6.24	9.91
TGDL-METIS	13.21	17.15	21.30	28.80	8.33	10.36	10.85	13.29	17.45	22.40	6.46	7.64
TGDL-DBH	12.82	16.35	21.07	27.30	8.27	10.09	10.31	13.18	16.59	22.20	6.19	7.60
TGDL	**12.45**	**15.87**	**20.00**	**26.37**	**8.11**	**9.92**	**9.75**	**12.45**	**15.97**	**21.46**	**5.68**	**6.98**

6.4.3 Ablation Study of Key Modules

To analyze the impact of proposed components on final prediction performance, ablation studies were conducted with five designed variants for comparison: (1) *w/o Structure Dec*: This variant removes the graph structure decomposition module, using the original graph directly as input structure for each ST block. (2) *w/o Signal Dec*: This variant removes the Residual Decomposition Flow described in Sect. 6.3.3. (3) *w/o Indep Reg*: This variant excludes all independence regularization terms, including constraint terms in Eqs. (6.67) and (6.78). (4) *w/o Structure Dec+Indep Reg*: This variant is the combination of variants (1) and (3). (5) *w/o Signal Dec+Indep Reg*: This variant is the combination of variants (2) and (3). The last two variants aim to examine the synergistic effects among different modules and constraints.

Experimental results using STGCN as the base model are listed in Table 6.5. As illustrated, each component contributes to performance improvement. Specifically, the significant performance drops in variants (1) and (2) on PEMSD datasets demonstrate that the proposed structure and signal decomposition methods are particularly critical when dealing with traffic data possessing complex component couplings. The performance decline in variant (3) further validates the effectiveness of independence constraints. Moreover, variants (4) and (5) exhibit a notable decline in performance

Table 6.5 Ablation study results of the TGDL framework

Dataset	NYCBike			NYCTaxi		
Metric	MAE ↓	RMSE ↓	MAPE ↓	MAE ↓	RMSE ↓	MAPE ↓
(0): TGDL	**5.22**	**7.66**	**22.52**	**10.63**	**18.88**	**17.07**
(1): w/o Structure Dec	5.28	<u>7.73</u>	<u>22.54</u>	<u>10.71</u>	<u>19.25</u>	17.36
(2): w/o Signal Dec	5.30	7.79	22.78	11.37	21.01	17.43
(3): w/o Indep Reg	<u>5.27</u>	7.75	22.57	11.40	21.34	<u>17.29</u>
(4): w/o Structure Dec+Indep Reg	5.45	8.05	23.14	11.94	21.76	17.61
(5): w/o Signal Dec+Indep Reg	5.89	8.46	25.77	12.65	23.13	19.30
Dataset	PEMSD4			PEMSD8		
Metric	MAE ↓	RMSE ↓	MAPE ↓	MAE ↓	RMSE ↓	MAPE ↓
(0): TGDL	**14.25**	**23.51**	**8.96**	**11.56**	**19.46**	**6.57**
(1): w/o Structure Dec	15.63	25.33	9.99	12.67	21.40	7.24
(2): w/o Signal Dec	<u>15.35</u>	<u>25.08</u>	9.82	13.13	21.67	7.49
(3): w/o Indep Reg	15.67	25.58	<u>9.80</u>	<u>12.63</u>	<u>21.13</u>	<u>7.22</u>
(4): w/o Structure Dec+Indep Reg	16.12	25.97	10.24	13.09	22.44	7.28
(5): w/o Signal Dec+Indep Reg	16.79	26.93	10.32	14.32	24.78	7.96

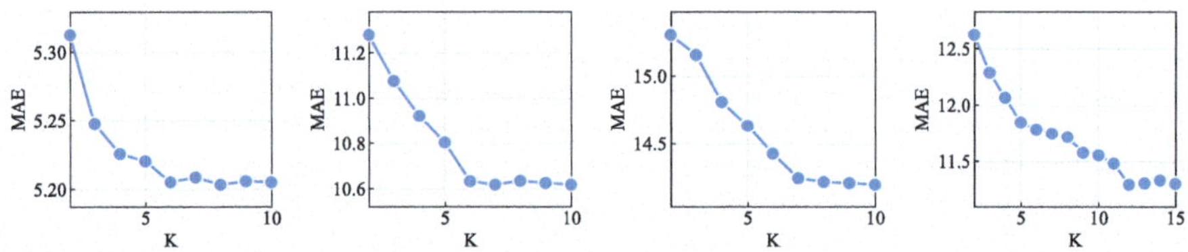

(a) NYCBike Dataset (b) NYCTaxi Dataset (c) PEMSD4 Dataset (d) PEMSD8 Dataset

Fig. 6.4 Hyperparameter experiment results for TGDL framework

compared to individual ablations, suggesting the presence of a synergistic effect among the proposed modules, i.e., the combined effect of multiple modules exceeds the sum of their individual contributions. Additionally, variant (5) performs worse than variant (4), indicating that the synergy between signal decomposition and the independence regularization term plays a particularly critical role in enhancing model performance.

In experiments, default configurations from official open-source code were used to ensure optimal base model performance. The only hyperparameter introduced by the TGDL framework is the number of decomposition sub-graphs K. To analyze the impact of this parameter, K was incrementally increased from 2 to 15 across all four datasets using STGCN as the base model.

As shown in Fig. 6.4 (note that for NYCBike, NYCTaxi, and PEMSD4, we removed results above 10 because they remained essentially unchanged), when K is too small, model performance drops significantly because such a configuration struggles to effectively capture the influence of multiple factors in traffic data. As K increases, performance gradually stabilizes and improves slightly in some cases. Based on experimental results, to achieve the optimal balance between model performance and complexity, K was set to 6, 6, 7, and 12 for NYCBike, NYCTaxi, PEMSD4, and PEMSD8 datasets, respectively.

6.4.4 *In-Depth Analysis of Model Behavior*

This subsection provides an in-depth examination of model behavior by analyzing the effect of independence regularization, visualizing learned signal representations, and evaluating model efficiency and scalability. These analyses offer insights into the theoretical validity, representational properties, and practical feasibility of the proposed framework.

Table 6.6 Comparison of mutual information between prediction results of different components

Dataset	NYCBike	NYCTaxi	PEMSD4	PEMSD8
TGDL w/o Indep Reg	0.2305	0.6197	0.2306	0.3616
TGDL	0.0957	0.1678	0.0603	0.0585
Relative reduction	−58.48%	−72.92%	−73.85%	−83.82%

6.4.4.1 Analysis of Independence Regularization

Based on the Component Independence (CI) Principle, the TGDL framework employs two independence regularization strategies to ensure mutual independence of signal components on different sub-graphs.

To verify the effect of independence regularization, experiments were conducted on the PEMSD8 dataset using TGDL based on STGCN, with correlation matrices visualized in Fig. 6.5. In this experiment, traffic states of the PEMSD8 dataset are decomposed into 12 sub-graphs. The correlation between different sub-graph components is represented by a correlation matrix, where the (k, k')-th element in Fig. 6.5 reflects the normalized mutual information between the k-th and k'-th sub-graphs. Values closer to 1 indicate higher correlation, and regions with larger correlation values appear brighter in the figure.

Figure 6.5a shows the correlation matrix without independence regularization, while Fig. 6.5b displays the result with independence regularization. Comparison reveals that independence regularization effectively reduces the correlation between sub-graph components, thereby promoting component decoupling. This satisfies the CI Principle, enabling TGDL to achieve predictions with lower DI error. Additionally, Table 6.6 reports results for other datasets, specifically statistics on the mean value of the upper triangular part of the correlation matrix. Experimental results indicate that the introduction of independence regularization reduces correlation by more than 50% across all dataset experiments, significantly decreasing DI error.

Notably, some non-zero inter-component correlations persist after decomposition, as shown in Fig. 6.5. This indicates that our regularization does not strictly remove all dependencies but rather reorganizes the modeling focus. It encourages the model to capture strong local dynamics explicitly while retaining the flexibility to model long-range or global interactions implicitly. This balance contributes to the robust performance of TGDL across diverse scenarios.

6.4.4.2 Visualization Analysis of Signal Representations

To further explore the role of independence regularization terms in TGDL, this experiment visually analyzes the representation distribution of the signal representation matrix $\mathbf{H}_t^{(k)}$. This matrix is generated by the signal decomposition process, calculated

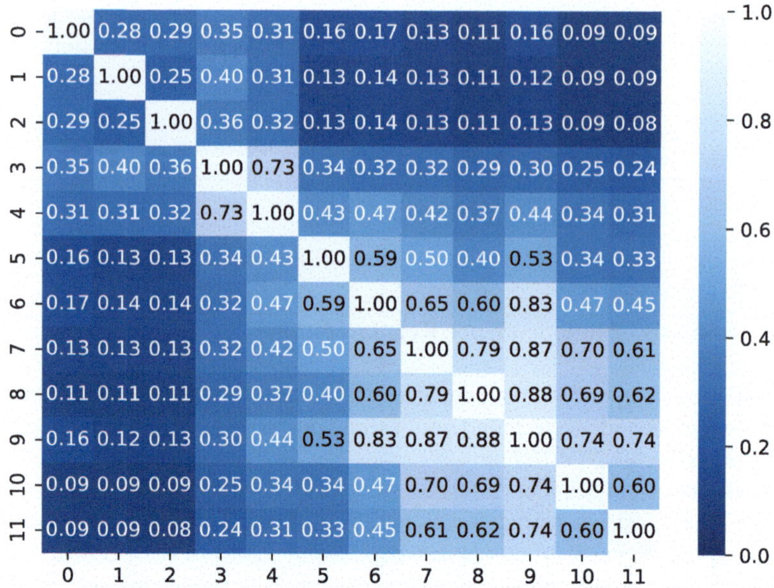

(a) TGDL without independence regularization

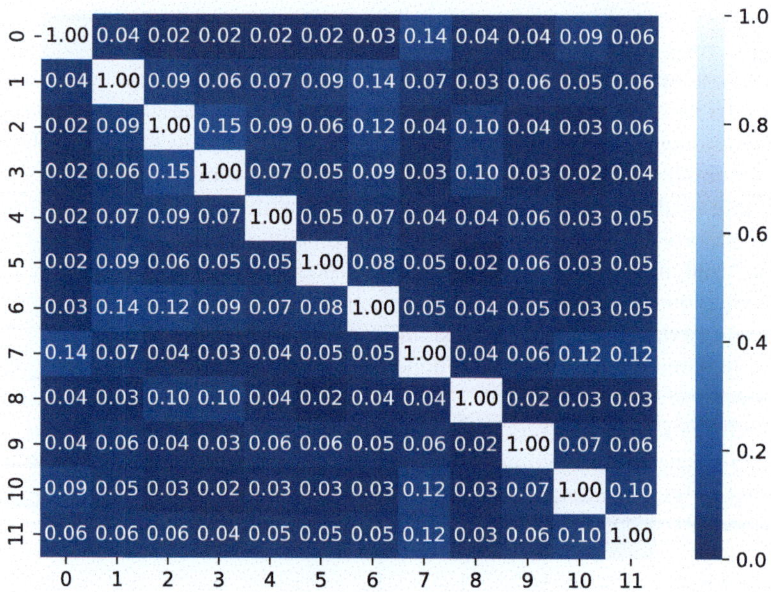

(b) TGDL

Fig. 6.5 Correlation matrix between prediction results of different components

(a) w/o TGDL (b) w/o training ($S = 0.27$) (c) w/o ind ($S = 0.68$) (d) TGDL ($S = 0.77$) (e) SCNN ($S = 0.50$)

Fig. 6.6 Visualization of learned representations for different models

as per Eq. (6.74), where $\mathbf{H}_t^{(k)}$ represents the signal representation of sub-graph k at time step t. In the experiment, representations $\mathbf{H}_t^{(k)}$ were generated for all samples in the PEMSD8 dataset, and the t-SNE algorithm [19] was used to reduce dimensionality to 2D to visualize distribution characteristics of different sub-graph signals.

Figure 6.6 shows representation distributions under different conditions to analyze the role of independence regularization in TGDL. Specifically, Fig. 6.6a displays the representation distribution of the base model (STGCN); Fig. 6.6b shows the untrained TGDL model; Fig. 6.6c shows the TGDL model without independence regularization; and Fig. 6.6d represents representations generated by the complete TGDL model. For comparison, Fig. 6.6e visualizes the representation distribution of the optimal signal decomposition prediction model, SCNN. In these visualizations, each point represents a sample, and colors distinguish representation distributions of different sub-graphs. To further quantify these distribution differences, the Silhouette Score [20] was calculated to evaluate clustering performance, denoted as "S" in the figure, where higher values indicate better performance.

From the visualization results in Fig. 6.6, significant differences in representation distributions among models can be observed. Representations generated by the base model are relatively scattered and lack clear clustering patterns, indicating limited capability to distinguish different sub-graph signals. The untrained TGDL, having not undergone training, shows representations that are somewhat separated but difficult to distinguish, suggesting the model has not yet learned effective sub-graph decoupling features. TGDL without independence regularization can distinguish different components reasonably well in most cases, but some representation overlap persists, indicating partial failure to fully decouple sub-graph signals. In contrast, the complete TGDL model significantly improves representation distribution, with all sub-graph signals clearly separated, validating the effectiveness of independence regularization for component decoupling.

Furthermore, although the SCNN model can distinguish different representations to some extent, it struggles to effectively cluster samples of the same sub-component, indicating limitations in its decoupling capability. Comparing Silhouette Coefficients between SCNN and TGDL models (and variants), it is evident that even TGDL without independence regularization achieves better decomposition than SCNN, further explaining TGDL's superior prediction performance over SCNN.

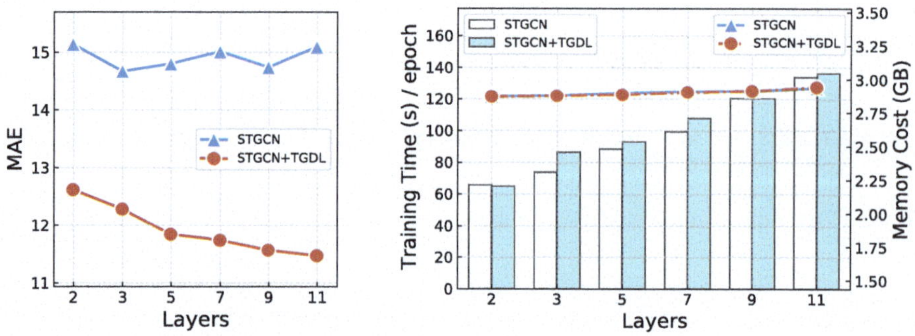

(a) Scalability Experiment Results (b) Time/Memory Cost

Fig. 6.7 Scalability and efficiency analysis of TGDL framework

6.4.4.3 Model Efficiency and Scalability Analysis

TGDL enhances traffic state prediction accuracy by decomposing the original traffic data into multiple components and predicting using multiple base models. However, since TGDL employs K base models, its parameter count increases K-fold compared to a single base model. A natural question arises: Is TGDL's performance improvement solely due to increased model parameters? To investigate this, scalability experiments were conducted to evaluate TGDL's performance, training time, and memory usage under different sub-graph counts ($K = 2$ to $K = 11$). Results are shown in Fig. 6.7. The base model used was STGCN, and the dataset was PEMSD8. For fair comparison, a control experiment was designed: a standard STGCN base model was stacked with 2 to 11 layers via residual connections to match TGDL's parameter scale. This design ensures STGCN expands only through parameter increase, allowing direct comparison of whether TGDL's performance gain at the same parameter scale stems from the model architecture itself rather than mere parameter growth.

Results in Fig. 6.7 show that as the number of sub-graphs increases, TGDL effectively utilizes additional parameters to improve prediction performance. In contrast, stacking STGCN layers does not yield significant performance gains, suggesting that STGCN may have already minimized model-induced MI error, nearing the modeling upper bound for the original data. However, TGDL achieves better prediction performance by decomposing original data into easier-to-predict sub-components, thereby reducing data-induced DI error. Regarding computational resources, TGDL and the stacked STGCN model exhibit similar trends. First, training time increases slightly with parameter growth but does not show exponential growth. Second, memory usage remains relatively stable, maintaining between 2–3 GB, indicating that TGDL maintains good computational efficiency while enhancing model capability.

Overall, TGDL demonstrates good scalability while enhancing base model prediction capability, with computational resource consumption consistent with base models, thus providing an efficient and scalable spatiotemporal forecasting framework.

6.5 Summary

This chapter established the theoretical foundation of the "Decompose-Then-Predict" paradigm based on information theory and proposed the TGDL framework.

First, the traffic prediction problem was analyzed based on noisy channel coding theory, identifying that prediction error stems from data error and model error, whereas existing prediction model optimization typically focuses on minimizing model error, neglecting the impact of data error.

Based on this, Fano's Inequality was used to derive sufficient conditions for effective decomposition methods capable of reducing data error, formalized as the Component Independence (CI) Principle. Furthermore, we theoretically proved the existence of such independent decomposition for complex traffic data, ensuring the feasibility of the proposed principle. Guided by the CI Principle, a model-agnostic Theory-guided Graph Decomposition Learning (TGDL) framework was proposed, comprising the following core modules: (1) *Graph Structure Decomposition Module*: Introduces structure independence regularization to decompose the input traffic graph into multiple independent sub-graphs; (2) *Graph Signal Decomposition Module*: Employs a Residual Decomposition Flow with signal independence regularization to progressively decouple sub-graph signals, achieving effective information separation; (3) *Traffic Prediction Integration Module*: Aggregates predicted signals from each sub-graph to generate the final traffic state forecast.

The TGDL framework generalizes existing spatiotemporal prediction models using an "Encoder-Decoder" architecture, incorporating encoder and decoder modules as interchangeable components to adapt to different spatiotemporal prediction models.

Extensive experiments on four traffic datasets validated the portability and effectiveness of TGDL. As a general framework, TGDL can be flexibly applied to various graph-based spatiotemporal prediction models, significantly enhancing their performance and breaking through the performance bottlenecks of mainstream models while maintaining high computational efficiency and scalability. Meanwhile, compared to decomposition prediction models lacking effective decomposition theoretical guarantees, TGDL achieves lower-error spatiotemporal prediction results. Furthermore, in-depth quantitative analysis of decomposition modules shows that experimental results align with the Component Independence Principle, further proving TGDL's effectiveness in modeling complex spatiotemporal relationships.

References

1. Beaudry, N.J., Renner, R.: An intuitive proof of the data processing inequality. Quantum Information and Computation **12**(5-6), 432–441 (2012)
2. Chen, C., Petty, K., Skabardonis, A., Varaiya, P., Jia, Z.: Freeway performance measurement system: Mining loop detector data. Transportation Research Record **1748**(1), 96–102 (2001)

3. Cheng, P., Hao, W., Dai, S., Liu, J., Gan, Z., Carin, L.: CLUB: A contrastive log-ratio upper bound of mutual information. In: Proceedings of the 37th International Conference on Machine Learning (ICML'20), pp. 1779–1788 (2020)
4. Cover, T.M., Thomas, J.A.: Elements of Information Theory. Wiley-Interscience (2006)
5. Deng, J., Chen, X., Jiang, R., Yin, D., Yang, Y., Song, X., Tsang, I.W.: Disentangling structured components: Towards adaptive, interpretable and scalable time series forecasting. IEEE Transactions on Knowledge and Data Engineering (TKDE) **36**(8), 3783–3800 (2024)
6. Fan, Z., Song, X., Shibasaki, R.: CitySpectrum: A non-negative tensor factorization approach. In: Proceedings of the 2014 ACM International Joint Conference on Pervasive and Ubiquitous Computing (UbiComp'14), pp. 213–223 (2014)
7. Fang, Y., Qin, Y., Luo, H., Zhao, F., Xu, B., Zeng, L., Wang, C.: When spatio-temporal meet wavelets: Disentangled traffic forecasting via efficient spectral graph attention networks. In: Proceedings of the 2023 IEEE 39th International Conference on Data Engineering (ICDE'23), pp. 517–529 (2023)
8. Fano, R.M., Hawkins, D.: Transmission of information: A statistical theory of communications. American Journal of Physics **29**(11), 793–794 (1961)
9. Geng, X., Li, Y., Wang, L., Zhang, L., Yang, Q., Ye, J., Liu, Y.: Spatiotemporal multi-graph convolution network for ride-hailing demand forecasting. In: Proceedings of the 33rd AAAI Conference on Artificial Intelligence (AAAI'19), pp. 3656–3663 (2019)
10. Guo, S., Lin, Y., Feng, N., Song, C., Wan, H.: Attention based spatial-temporal graph convolutional networks for traffic flow forecasting. In: Proceedings of the 33rd AAAI Conference on Artificial Intelligence (AAAI'19), pp. 922–929 (2019)
11. He, S., Ji, J., Lei, M.: Decomposed spatio-temporal mamba for long-term traffic prediction. In: Proc. of AAAI, pp. 11772–11780 (2025)
12. Ji, J., Wang, J., Huang, C., Wu, J., Xu, B., Wu, Z., Junbo, Z., Zheng, Y.: Spatio-temporal self-supervised learning for traffic flow prediction. In: Proceedings of the 37th AAAI Conference on Artificial Intelligence (AAAI'23), pp. 4356–4364 (2023)
13. Karypis, G., Kumar, V.: METIS: A software package for partitioning unstructured graphs, partitioning meshes, and computing fill-reducing orderings of sparse matrices. Tech. rep., University of Minnesota, Department of Computer Science and Engineering (1997)
14. Li, F., Yan, H., Jin, G., Liu, Y., Li, Y., Jin, D.: Automated spatio-temporal synchronous modeling with multiple graphs for traffic prediction. In: Proceedings of the 31st ACM International Conference on Information and Knowledge Management (CIKM'22), pp. 1084–1093 (2022)
15. Li, Y., Yu, R., Shahabi, C., Liu, Y.: Diffusion convolutional recurrent neural network: Data-driven traffic forecasting. In: Proceedings of the 6th International Conference on Learning Representations (ICLR'18), pp. 1–16 (2018)
16. Liu, D., Wang, J., Shang, S., Han, P.: MSDR: Multi-step dependency relation networks for spatial temporal forecasting. In: Proceedings of the 28th ACM SIGKDD Conference on Knowledge Discovery and Data Mining (KDD'22), pp. 1042–1050 (2022)
17. Liu, S., Johns, E., Davison, A.J.: End-to-end multi-task learning with attention. In: Proceedings of the 2019 IEEE Conference on Computer Vision and Pattern Recognition (CVPR'19), pp. 1871–1880 (2019)
18. Lu, W., Hu, Y., Chen, W., Qin, Y., Wu, C., He, X., et al.: Traffic flow prediction for highway vehicle detectors through decomposition and machine learning. Transportation Letters **17**(2), 260–280 (2025)
19. Maaten, L.V.D., Hinton, G.: Visualizing data using t-SNE. Journal of Machine Learning Research (JMLR) **9**(86), 2579–2605 (2008)
20. Rousseeuw, P.J.: Silhouettes: A graphical aid to the interpretation and validation of cluster analysis. Journal of Computational and Applied Mathematics **20**(1), 53–65 (1987)
21. Shao, Z., Zhang, Z., Wei, W., Wang, F., Xu, Y., Cao, X., Jensen, C.S.: Decoupled dynamic spatial-temporal graph neural network for traffic forecasting. Proceedings of the VLDB Endowment **15**(11), 2733–2746 (2022)
22. Soltani, S.: On the use of the wavelet decomposition for time series prediction. Neurocomputing **48**(1), 267–277 (2002)

23. Taylor, J.E., Rousselet, G.A., Sereno, S.C.: Can prediction error explain predictability effects on the N1 during picture-word verification? Imaging Neuroscience **2**(1), 1–24 (2024)
24. Wang, J., Ji, J., Jiang, Z., Sun, L.: Traffic flow prediction based on spatiotemporal potential energy fields. IEEE Transactions on Knowledge and Data Engineering (TKDE) **35**(9), 9073–9087 (2023)
25. Wang, J., Mao, Y., Li, J., Xiong, Z., Wang, W.X.: Predictability of road traffic and congestion in urban areas. PloS one **10**(4), e0121825 (2015)
26. Wang, J., Wang, Z., Li, J., Wu, J.: Multilevel wavelet decomposition network for interpretable time series analysis. In: Proceedings of the 24th ACM SIGKDD Conference on Knowledge Discovery and Data Mining (KDD'18), pp. 2437–2446 (2018)
27. Wang, J., Wu, J., Wang, Z., Gao, F., Xiong, Z.: Understanding urban dynamics via context-aware tensor factorization with neighboring regularization. IEEE Transactions on Knowledge and Data Engineering (TKDE) **32**(11), 2269–2283 (2019)
28. Woo, G., Liu, C., Sahoo, D., Kumar, A., Hoi, S.: CoST: Contrastive learning of disentangled seasonal-trend representations for time series forecasting. In: Proceedings of the 10th International Conference on Learning Representations (ICLR'22), pp. 1–18 (2022)
29. Wu, Z., Pan, S., Long, G., Jiang, J., Chang, X., Zhang, C.: Connecting the dots: Multivariate time series forecasting with graph neural networks. In: Proceedings of the 26th ACM SIGKDD Conference on Knowledge Discovery and Data Mining (KDD'20), pp. 753–763 (2020)
30. Wu, Z., Pan, S., Long, G., Jiang, J., Zhang, C.: Graph WaveNet for deep spatial-temporal graph modeling. In: Proceedings of the 28th International Joint Conference on Artificial Intelligence (IJCAI'19), pp. 1907–1913 (2019)
31. Xie, C., Yan, L., Li, W., Zhang, Z.: Distributed power-law graph computing: Theoretical and empirical analysis. In: Proceedings of the 28th International Conference on Neural Information Processing Systems (NeurIPS'14), pp. 1673–1681 (2014)
32. Yao, H., Tang, X., Wei, H., Zheng, G., Li, Z.: Revisiting spatial-temporal similarity: A deep learning framework for traffic prediction. In: Proceedings of the 33rd AAAI Conference on Artificial Intelligence (AAAI'19), pp. 5668–5675 (2019)
33. Yu, B., Yin, H., Zhu, Z.: Spatio-temporal graph convolutional networks: A deep learning framework for traffic forecasting. In: Proceedings of the 27th International Joint Conference on Artificial Intelligence (IJCAI'18), pp. 3634–3640 (2018)
34. Zhang, J., Zheng, Y., Qi, D.: Deep spatio-temporal residual networks for citywide crowd flows prediction. In: Proceedings of the 31st AAAI Conference on Artificial Intelligence (AAAI'17), pp. 1655–1661 (2017)
35. Zhou, W., Wang, W., Hua, X., Zhang, Y.: Real-time traffic flow forecasting via a novel method combining periodic-trend decomposition. Sustainability **12**(15), 5891 (2020)

Chapter 7
Conclusion and Future Perspectives

Abstract This final chapter summarizes the main contributions and empirical insights presented throughout the book and outlines promising directions for future research. Centered on mechanism-driven and explainable spatio-temporal prediction in urban environments, the book has systematically investigated how domain mechanisms can be integrated into deep learning models to enhance interpretability, robustness, and generalization under complex and dynamic conditions. Building on these findings, this chapter reflects on remaining challenges and discusses future perspectives on data, methodology, and application domains, aiming to inform and inspire subsequent research in hybrid spatio-temporal modeling.

7.1 Summary of Contributions and Key Findings

This book is devoted to the study of mechanism-driven and explainable urban spatio-temporal prediction, with urban traffic systems serving as the primary application domain. From two key perspectives, i.e. urban population mobility mechanisms and scenario-driven environmental dynamics, it systematically investigates the fundamental challenges in spatio-temporal forecasting. In response to the limitations of existing methods, such as insufficient theoretical grounding and limited generalization ability in traffic prediction tasks, this book develops a series of innovative approaches that address core scientific problems, including discrete-time potential field modeling, continuous-time potential field dynamics, spatio-temporal distribution shift modeling, and theory-guided spatio-temporal decomposition. Together, these contributions provide new perspectives and effective solutions for mechanism-driven and explainable spatio-temporal prediction.

The main contributions and findings of this book can be summarized as follows.

First, inspired by the gravity theory of human mobility, the book introduces a spatio-temporal potential energy field extraction and prediction model, termed STPEF, to explain the evolution of regional traffic flows. Given regional traffic flow data, STPEF extracts interpretable potential energy fields that capture underlying population movement mechanisms and leverages them for future traffic prediction. Experimental results on regional traffic flow datasets from three cities demonstrate

J. Wang and J. Ji, *Mechanism-Driven Explainable Urban Spatio-Temporal Prediction*, Big Data Management, https://doi.org/10.1007/978-981-92-0662-9_7

that STPEF successfully learns meaningful potential fields consistent with mobility mechanisms, while achieving prediction performance comparable to, and in some cases better than, existing methods.

Second, guided by the potential field formulation and the continuity equation derived from energy conservation principles, the book develops a physics-informed spatio-temporal differential equation network, STDEN. By introducing graph neural networks with continuous hidden layers, STDEN models the spatial evolution of traffic potential fields over continuous time and generates predictions that satisfy physical constraints of road network dynamics. Empirical evaluations on road network traffic datasets from multiple functional zones show that STDEN consistently outperforms deep learning models that ignore temporal continuity in traffic flow modeling.

Third, grounded in causal modeling and backdoor adjustment theory, the book proposes a confounder-aware spatio-temporal basis vector representation and prediction model, STEVE. By adopting a basis vector formulation, STEVE removes the restrictive assumption of discrete and predefined confounders, enabling the representation of arbitrary and unknown confounding factors. This design allows the model to disentangle confounding effects and accurately capture their influence on future data distributions, thereby improving robustness and generalization. The effectiveness of STEVE is validated on four datasets involving both temporal and spatial distribution shifts.

Finally, drawing on the noisy channel model and Fano's inequality from information theory, the book establishes theoretical conditions under which decomposition algorithms can effectively reduce data complexity. Based on this analysis, a model-agnostic spatio-temporal decomposition prediction framework, TGDL, is proposed. By decomposing complex spatio-temporal signals into multiple sub-components under theoretical guidance, TGDL significantly reduces modeling complexity and enhances prediction performance in challenging traffic scenarios. Extensive experiments on four real-world traffic datasets demonstrate that TGDL successfully breaks the performance bottlenecks of several mainstream spatio-temporal prediction models, achieving superior results with theoretical guarantees compared to existing decomposition-based approaches.

7.2 Future Research Directions and Open Challenges

Scientific research paradigms can broadly be categorized into two complementary types. The first is the Keplerian, data-driven paradigm, which relies on large-scale data and artificial intelligence models to uncover underlying patterns. The second is the Newtonian, mechanism-driven paradigm, which starts from first principles to derive theoretical models and explanatory frameworks.

In the field of spatio-temporal prediction, purely data-driven deep learning models have been extensively explored and have achieved remarkable empirical success. However, as research progresses, their inherent limitations have become increasingly evident. In particular, such models often overlook the intrinsic mechanisms governing data generation, leading to limited interpretability of prediction results. Moreover, the absence of explicit mechanism modeling tends to impair their generalization ability when confronted with dynamic and complex distributional shifts.

To address these challenges, this book explores the integration of data-driven and mechanism-driven paradigms by injecting mechanistic knowledge into deep spatio-temporal prediction models. The objective is to enhance their capability to capture the internal evolutionary processes of spatio-temporal data and to improve generalization across diverse scenarios. Although meaningful progress has been made, several limitations remain, which also point to promising directions for future research. These perspectives are discussed below from the viewpoints of data, methodology, and application domains.

From the *data* perspective, the studies in this book primarily rely on existing traffic state data for modeling and prediction. While spatio-temporal dependencies are explicitly considered, the reliance on a single data modality limits both model generalization and adaptability to complex traffic environments. In reality, traffic states are influenced not only by historical flow patterns but also by a wide range of external factors, such as road infrastructure conditions, weather variations, and traffic incidents. Although this book attempts to implicitly capture such effects from traffic data, it does not explicitly incorporate heterogeneous data sources. This is partly due to the difficulty of acquiring external data and partly because these factors are inherently embedded in the data generation process. In addition, individual-level traffic behaviors, including travel patterns and driving habits, play a critical role in shaping aggregate traffic dynamics. However, most existing studies, including those in this book, focus on macroscopic flow modeling and underutilize fine-grained behavioral data. Future work may enrich data sources by integrating multi-source heterogeneous data, such as congestion reports, weather forecasts, and holiday travel patterns, to improve robustness to environmental changes. Furthermore, incorporating individual-level data, such as GPS trajectories and travel mode information, could enable finer-grained modeling of traffic dynamics.

From the *methodological* perspective, the proposed approaches aim to effectively integrate scenario-specific mechanistic models into deep learning frameworks to enhance predictive performance and interpretability. Nevertheless, mechanistic models are often domain- and scenario-dependent, requiring careful adaptation across different applications. In recent years, large-scale models, especially foundation models, have demonstrated strong generalization and adaptability across a wide range of tasks, emerging as a central direction in artificial intelligence research. Future studies may investigate how cross-domain and cross-scenario mechanistic knowledge can be embedded into the training of spatio-temporal foundation models, thereby enabling the development of explainable large-scale spatio-temporal models with built-in physical or causal constraints. Such models could potentially generalize beyond spatio-temporal prediction to a broader class of spatio-temporal understanding and reasoning tasks.

From the *application* perspective, this book focuses primarily on urban traffic systems. While the proposed methods show strong performance in this domain, their applicability to other spatio-temporal fields remains to be further validated. On the one hand, spatio-temporal data exhibit distinct characteristics across domains: traffic flow data are typically represented as directed network dynamics, whereas other domains, such as air quality monitoring or epidemic spread, involve different spatial dependencies and propagation mechanisms. On the other hand, application scenarios impose varying requirements on prediction accuracy, interpretability, and real-time responsiveness, which are not explicitly optimized in the current approaches. Future research may extend the proposed framework to a broader range of spatio-temporal prediction tasks, including air pollution diffusion modeling and extreme weather forecasting, to assess its generality and adaptability. Moreover, different domains may necessitate tailored modeling strategies. For example, epidemic prediction emphasizes individual behaviors and social network diffusion, whereas weather forecasting requires more detailed physical modeling. Consequently, further exploration is needed to adapt mechanism-driven strategies and model architectures to domain-specific requirements, with the goal of achieving optimal performance across diverse spatio-temporal prediction scenarios.

The manufacturer's authorised representative in the EU is Springer Nature Customer Service Centre GmbH, Europaplatz 3, 69115 Heidelberg, Germany. If you have any concerns regarding our products, please contact ProductSafety@springernature.com

Printed and bound by CPI Group (UK) Ltd, Croydon, CR0 4YY
21/09/2026
02219993-0015